T0339904

Means in Mathematical Analysis

Mathematical Analysis and its Applications Series

Means in Mathematical Analysis

Bivariate Means

Gheorghe Toader
Technical University of Cluj-Napoca, Cluj-Napoca, Romania

Iulia Costin
Technical University of Cluj-Napoca, Cluj-Napoca, Romania

Series Editor
Themistocles M. Rassias

Academic Press is an imprint of Elsevier
125 London Wall, London EC2Y 5AS, United Kingdom
525 B Street, Suite 1800, San Diego, CA 92101-4495, United States
50 Hampshire Street, 5th Floor, Cambridge, MA 02139, United States
The Boulevard, Langford Lane, Kidlington, Oxford OX5 1GB, United Kingdom

Notices

Knowledge and best practice in this field are constantly changing. As new research and experience broaden our understanding, changes in research methods, professional practices, or medical treatment may become necessary.

Practitioners and researchers must always rely on their own experience and knowledge in evaluating and using any information, methods, compounds, or experiments described herein. In using such information or methods they should be mindful of their own safety and the safety of others, including parties for whom they have a professional responsibility.

To the fullest extent of the law, neither the Publisher nor the authors, contributors, or editors, assume any liability for any injury and/or damage to persons or property as a matter of products liability, negligence or otherwise, or from any use or operation of any methods, products, instructions, or ideas contained in the material herein.

Library of Congress Cataloging-in-Publication Data
A catalog record for this book is available from the Library of Congress

British Library Cataloguing-in-Publication Data
A catalogue record for this book is available from the British Library

ISBN: 978-0-12-811080-5

For information on all Academic Press publications
visit our website at https://www.elsevier.com/books-and-journals

Publisher: Candice Janco
Senior Acquisitions Editor: Graham Nisbet
Editorial Project Manager: Susan Ikeda
Production Project Manager: Poulouse Joseph
Designer: Alan Studholme

Typeset by VTeX

This book is dedicated to the loving memory of my father, Professor Gheorghe Toader, and it crowns his career as a tireless mathematician and professor. It also is dedicated to my mother, Associate professor Silvia Toader, who urged us along the way to finish it.

Iulia Costin

Contents

About the Authors

Gheorghe Toader (1945–2015) graduated in 1969 from the Faculty of Mathematics at "Babeş–Bolyai" University in Cluj-Napoca. In 1980 he defended his PhD thesis. His last position was as a Professor at the Mathematics Department of Technical University of Cluj-Napoca. He has published over 165 scientific papers in the field of Mathematics, of which more than 70 papers related to the subject of this book. He has been a reviewer of Zentralblatt für Mathematik (Springer, Germany) and since 2007 he was in the Editorial Board of Journal of Mathematical Inequalities (Zagreb, Croatia). He was also a member of Research Group on Mathematical Inequalities and Applications (Melbourne, Australia), of the Romanian Mathematical Society, and of the Mathematical Foundation "Th. Angheluţă" (Cluj-Napoca, Romania).

Iulia Costin was born in Romania in 1971. She graduated in 1994 from the Faculty of Mathematics and Informatics at "Babeş–Bolyai" University in Cluj-Napoca. In 2001 she defended her PhD thesis. Currently she is a Senior Lecturer at the Computer Science Department of Technical University of Cluj-Napoca. She has published more than 30 papers related to the subject of this book. She is a former member of the IEEE Computer Society, and since 2005 is a member of the ACM (Association for Computer Machinery).

Preface

The subjects of interest in this book – means and double sequences – are generally associated with the names of great mathematicians such as Pythagoras, Archimedes, Heron, J.L. Lagrange, and C.F. Gauss, to list but a few of the more famous names.

With this book we continue the subject developed in some of our previous papers. Some computations needed for obtaining some of our results were performed using the computer algebra program Maple.

We hope that this book clearly addresses the modern possibilities of developing an old subject such as the one given by C.F. Gauss on double sequences, assisting those interested in the history of the subject, as well as researchers willing to go further in some of the indicated directions.

Iulia Costin
Cluj-Napoca, Romania
April, 2017

Acknowledgment

We thank Professor Themistocles M. Rassias, who kindly accepted to publish our book, and the anonymous referees for their careful reviewing, for their observations and for their positive appreciations. We also thank Susan E. Ikeda and Graham Nisbet at Elsevier for their very useful and patient guidance throughout the publishing process.

Iulia Costin
April, 2017

Introduction

Motivation for this book

The subjects of interest in this book – means and double sequences – are generally associated with the names of great mathematicians such as Pythagoras, Archimedes, Heron, J.L. Lagrange , and C.F. Gauss, to list but a few of the more famous names. Therefore we have to begin with a short introduction into the history of mathematics. Passing then to the current stage of development in the field of means and double sequences, we, as the authors, will refer to our own results, most of which have been published in Romanian journals with limited distribution and may therefore be less known. This doesn't mean that even one relevant result known to us will be omitted.

Throughout the book, references to papers and books are indicated with the name of the author (authors) followed by the year of publication. If an author (or a group of authors) is (are) present in the list of references with more papers published in the same year, small letters are appended to the corresponding year.

With this book we continue the subject developed in Toader and Toader (2005), passing from Greek means to arbitrary means. By Greek means we refer to the means defined by the Pythagorean school. These were presented by Pappus of Alexandria in his books in the fourth century AD (see Pappus, 1932) and they include six (unnamed) means, along with the four well-known means:

– the arithmetic mean $\mathcal{A}$, defined by

$$\mathcal{A}(a,b)=\frac{a+b}{2},\ a,b>0;$$

– the geometric mean $\mathcal{G}$, given by

$$\mathcal{G}(a,b)=\sqrt{ab},\ a,b>0;$$

– the harmonic mean $\mathcal{H}$, with the expression

$$\mathcal{H}(a,b)=\frac{2ab}{a+b},\ a,b>0;$$

– the contraharmonic mean C, defined by

$$C(a,b) = \frac{a^2+b^2}{a+b}, \; a, b > 0.$$

Chapter 1

Chapter 1 of the book presents some classical problems involving double sequences. The oldest problem involving double sequences was given by Archimedes of Syracuse (287–212 BC) in his book *Measurement of the Circle*. The problem he was trying to solve was the evaluation of the number π, defined as the ratio of the perimeter of a circle to its diameter.

Consider a circle of radius 1. Denoting by p_n and P_n the half of the perimeters of the inscribed and circumscribed regular polygons with n sides, respectively, we have

$$p_n < \pi < P_n, \; n \geq 3.$$

To obtain a good estimation, Archimedes passes from a given n to $2n$, proving his famous inequalities

$$3.1408 < 3\frac{10}{71} < p_{96} < \pi < P_{96} < 3\frac{1}{7} < 3.1429.$$

As it was shown in Phillips (1981), the procedure can be presented as a double sequence. Denoting $P_{2^k n} = a_k$ and $p_{2^k n} = b_k$, the sequences $(a_k)_{k\geq 0}$ and $(b_k)_{k\geq 0}$ are given, step by step, by the relations

$$a_{k+1} = \mathcal{H}(a_k, b_k), b_{k+1} = \mathcal{G}(a_{k+1}, b_k), k \geq 0,$$

for some initial values a_0 and b_0. These sequences are monotonously convergent to a common limit, which, for $0 < b_0 < a_0$, has the value

$$\frac{a_0 b_0}{\sqrt{a_0^2 - b_0^2}} \arccos \frac{b_0}{a_0}.$$

In Archimedes' case

$$a_0 = P_3 = 3\sqrt{3} \text{ and } b_0 = p_3 = 3\sqrt{3}/2,$$

thus the common limit is π. In Borwein and Borwein (1983a, 1984), the authors describe another form of a double sequence (related also to the Gaussian arithmetic–geometric mean iteration), that is quadratically convergent to π. Also, Newman (1985) defines an alternate version of the arithmetic–geometric mean.

A second example of double sequences is provided by Heron's method of extracting square roots (as it is given in Bullen, 2003). To compute the geometric root of two numbers a and b, Heron used the arithmetic mean together with the harmonic mean. Letting $a_0 = a$ and $b_0 = b$, we can define

$$a_{k+1} = \mathcal{H}(a_k, b_k), b_{k+1} = \mathcal{A}(a_k, b_k), k \geq 0.$$

It is proven that the sequences $(a_k)_{k\geq 0}$ and $(b_k)_{k\geq 0}$ are monotonously convergent to the common limit $\sqrt{ab}$. Of course, the procedure was used only as an approximation method, the notion of limit being unknown in Heron's time.

The third example is Lagrange's method of determining certain irrational integrals. In Lagrange (1784–1985), for the evaluation of an integral of the form

$$\int \frac{N(y^2)dy}{\sqrt{(1+p^2y^2)(1+q^2y^2)}},$$

where N is a rational function, an iterative method is defined which leads to the rationalization of the function. As we shall see, a double sequence was defined by $a_0 = p$, $b_0 = q$, and

$$a_{k+1} = \mathcal{A}(a_k, b_k), b_{k+1} = \mathcal{G}(a_k, b_k), k \geq 0.$$

The same double sequence was considered in Gauss (1800), but only published many years later. For $a_0 = \sqrt{2}$ and $b_0 = 1$, Gauss remarked that a_4 and b_4 have the same first eleven decimals as those determined in Euler (1782) for the integral

$$2\int_0^1 \frac{z^2dz}{\sqrt{1-z^4}}.$$

In fact the sequences $(a_k)_{k\geq 0}$ and $(b_k)_{k\geq 0}$ are monotonously convergent to a common limit, which is now referred to as the arithmetic–geometric meanarithmetic–geometric mean ($\mathcal{AGM}$ for short) and it is denoted by $\mathcal{A} \otimes \mathcal{G}(a,b)$. Later, Gauss was able to represent the arithmetic–geometric mean using an elliptic integral, by

$$\mathcal{A} \otimes \mathcal{G}(a,b) = \frac{\pi}{2} \cdot \left[\int_0^{\pi/2} \frac{d\theta}{\sqrt{a^2\cos^2\theta + b^2\sin^2\theta}}\right]^{-1}.$$

As it is well known, this result is used for numerical evaluation of the elliptic integral.

The arithmetic–geometric mean or some similar defined means were studied in many other papers that were not explicitly mentioned in the book, like Borchardt (1861, 1876), Ciorănescu (1936), Cox (1985), Myrberg (1958a), Borwein and Borwein (1989), Schoenberg (1978) and many others.

Chapter 2

Chapter 2 is devoted mainly to those aspects of the theory of means needed in the study of double sequences. We consider some methods of constructing means, some inequalities between means, and we study some operations with means (as, for example, in Toader and Toader, 2006a). For the study of double sequences (which will follow in the next chapter), the most important notions are those of invariance and of complementariness of means. If

$$P(M, N) = P,$$

we call the mean P to be (M, N)-invariant and the mean N to be the complementary of the mean M with respect to the mean P. In Toader and Toader (2005), by direct computation, there were determined the ninety complementaries of a Greek mean with respect to another. For other special means, some problems of invariance were solved as functional equations. Here, in order to determine the complementary of an arbitrary mean with respect to another in a given family of means, we use the series expansion of means. The computations were performed using the computer algebra program Maple, as it is presented in Heck (2003). Many results regarding the complementariness, for example with respect to weighted Gini means or Stolarsky means, cannot be obtained without the help of computer algebra. We have to solve complicated systems of nonlinear equations, sometimes printed on more pages. Though obtained with the help of a computer, the results can be easily verified. Since the complementary of a mean with respect to another can be a pre-mean (thus a reflexive function), we transpose some properties from means to pre-means. We also study the problem of invariance in some families of means. Thus we determine some triples of means with the property that one of them is invariant with respect to the other two.

Chapter 3

In Chapter 3 of the book we focus on the general definition of double sequences. Namely, the Greek means, which appear in the examples from Chapter 1, are replaced by arbitrary means M and N. Before doing this, we have to remark that the essential difference between the first example and the other examples is the use of a_{k+1} in the definition of b_{k+1}. Therefore we can define double sequences of type

$$a_{k+1} = M(a_k, b_k), b_{k+1} = N(a_{k+1}, b_k), k \geq 0,$$

which are referred to as Archimedean double sequences in Phillips (1981), or of the type

$$a_{k+1} = M(a_k, b_k), b_{k+1} = N(a_k, b_k), k \geq 0,$$

referred to as Gaussian double sequences in Foster and Phillips (1984).

Related to a given Gaussian double sequence, the first problem which was studied is that of the convergence of the sequences $(a_k)_{k\geq 0}$ and $(b_k)_{k\geq 0}$ to a common limit. In this case it is proven that the common limit defines a mean, denoted by $M \otimes N$, and called the compound mean of M and N. There are many papers which deal with this problem, for example Schering (1878), Frisby (1879), Hettner (1880), Stieltjes (1880), Fricke (1901–1921), von Dávid (1907, 1909, 1913, 1928), Gosiewski (1909), Tonelli (1910), Kämmerer (1925), Geppert (1928, 1932, 1933), von Bültzingslöven (1933), Barna (1934, 1939), Mohr (1935), Aumann (1935), Ciorănescu (1936a), Uspensky (1945), Faragó (1951), Tietze (1952–1953), Stöhr (1956), Andreoli (1957a), Veinger (1964), Zhuravskii (1964), Gatteschi (1966, 1982), Rosenberg (1966), Allasia (1969–1970, 1970–1971, 1971–1972, 1983), Kuznecov (1969), Fuchs (1972), Heinrich (1981), Arazy et al. (1985), Reyssat (1987), Toader (1987b), Lohnsein (1888, 1888a), Grayson (1989), Peetre (1989, 1991), Claesson and Peetre (1990), Bullett (1991), Sándor (1994). In Foster and Phillips (1985), Toader (1990, 1991), and Costin and Toader (2004, 2004a) one looks after minimal conditions on the means M and N that assure that their compound mean exists. Using Gauss composition, Matkowski (2004) solved some problem posed by Th.M. Rassias.

The second main problem related to a given Gaussian double sequence is that of the determination of $M \otimes N$. Using Gauss' idea by which he was able to give the representation of $\mathcal{A} \otimes \mathcal{G}$, in Borwein and Borwein (1987) the following Invariance Principle is proven: if $M \otimes N$ exists and is continuous, then it is the unique mean P which is (M, N)-invariant. Therefore all the results related to invariant means, or equivalently to complementary means, permit the construction of a double sequence with known limit.

Finally, the rate of convergence of the sequences to the common limit is also studied. In our cases the rate of convergence is quadratic (or even faster as it is suggested by the numerical examples).

Chapter 4

The last chapter of the book is related to integral means. Keeping in mind the representation of the $\mathcal{AGM}$ by an (elliptic) integral mean, in Haruki (1991) the author started the study of some integral means of the same nature. They were extended in many other papers, arriving at the following expression in Toader and Toader (2002a):

$$M_{p,q,r}(a,b) = r^{-1} \circ p^{-1}\left[\frac{1}{2\pi}\int_0^{2\pi} p \circ q^{-1}\left(q \circ r(a)\cdot\cos^2\theta + q \circ r(b)\cdot\sin^2\theta\right)d\theta\right],$$

where p, q, r are adequate functions. A recurrence formula and a functional equation of Gaussian type are studied. They provide the ability to determine some cases in which $M_{p,q,r}$ is a given mean.

We hope that this book clearly addresses the modern possibilities of developing an old subject such as the one given by C.F. Gauss on double sequences, assisting those interested in the history of the subject, as well as researchers willing to go further in some of the indicated directions.

Chapter 1

Classical theory of the $\mathcal{AGM}$

ABSTRACT

We begin by presenting some classical examples of double sequences. They are related to the method of Archimedes for evaluating π, Heron's method of extracting square roots, Lagrange's procedure of determining the integral of some irrational functions, and Gauss' approximation of some elliptic integrals. For the definition of these double sequences, the arithmetic mean, the geometric mean, and the harmonic mean are used. Other examples of double sequences are related to lemniscatic integrals, of which some evaluations Gauss-tied to the study of $\mathcal{AGM}$, Gauss' ordinary hypergeometric series that gives another representation of $\mathcal{AGM}$, and Landen's transformation that can be applied to hypergeometric series. Also, the perimeter of an ellipse can be related to elliptic integrals and hypergeometric series.

1.1 MEASUREMENT OF THE CIRCLE

The undisputed leader of Greek scientists was Archimedes of Syracuse (287–212 BC). He is well known for many discoveries and inventions in physics and engineering, but also for his contributions to mathematics. One of these contributions is the evaluation of π, developed especially in his book *Measurement of the Circle*. For its presentation we will use the paper of Phillips (1981) and the book of Phillips (2000). We can refer also to the paper of Baxter (1981) and the book of Berggren et al. (1997).

As it is well known, π was defined as the ratio of the perimeter of a given circle to its diameter. Consider a circle of radius 1. Let p_n and P_n denote half of the perimeters of the inscribed and circumscribed regular polygons with n sides respectively. As Archimedes remarked, for every $n \geq 3$,

$$p_n < \pi < P_n.$$

To get an estimation with any accuracy, Archimedes passed from a given n to $2n$. By simple geometrical considerations, he obtained the relations

$$p_{2n}^2 = \frac{2np_n^2}{n + \sqrt{n^2 - p_n^2}}$$

Means in Mathematical Analysis. http://dx.doi.org/10.1016/B978-0-12-811080-5.00001-3

and

$$P_{2n} = \frac{2n P_n}{n + \sqrt{n^2 + P_n^2}}.$$

Beginning with inscribed and circumscribed regular hexagons, with $p_6 = 3$ and $P_6 = 2\sqrt{3}$, then using the above formulas, he proved his famous inequalities

$$3.1408 < 3\frac{10}{71} < p_{96} < \pi < P_{96} < 3\frac{1}{7} < 3.1429.$$

In Phillips (1981) and Phillips (2000), Archimedes method is developed differently. As

$$p_n = n \cdot \sin\frac{\pi}{n} \text{ and } P_n = n \cdot \tan\frac{\pi}{n},$$

it follows that

$$\begin{aligned} p_n + P_n &= n \cdot \sin\frac{\pi}{n} \cdot \left(1 + \frac{1}{\cos\frac{\pi}{n}}\right) = 2n \cdot \tan\frac{\pi}{n} \cdot \cos^2\frac{\pi}{2n} \\ &= 2 \cdot p_n \cdot P_n \cdot \frac{\cos^2\frac{\pi}{2n}}{2n \cdot \sin\frac{\pi}{2n} \cdot \cos\frac{\pi}{2n}} = \frac{2 \cdot p_n \cdot P_n}{P_{2n}}, \end{aligned}$$

and therefore

$$P_{2n} = \frac{2 \cdot p_n \cdot P_n}{p_n + P_n}.$$

Other simple calculations yield

$$p_n \cdot P_{2n} = 2n^2 \cdot \sin\frac{\pi}{n} \cdot \tan\frac{\pi}{2n} = 4n^2 \cdot \sin^2\frac{\pi}{2n} = p_{2n}^2,$$

or

$$p_{2n} = \sqrt{p_n \cdot P_{2n}}.$$

Thus

$$P_{2n} = \mathcal{H}(P_n, p_n) \text{ and } p_{2n} = \mathcal{G}(P_{2n}, p_n)$$

where $\mathcal{H}$ and $\mathcal{G}$ denote the harmonic and the geometric mean defined in the Introduction.

Let us renounce the geometrical origins of the terms P_n and p_n, and replace the sequences $(P_{2^n \cdot 3})_{n \geq 0}$ and $(p_{2^n \cdot 3})_{n \geq 0}$ by two positive sequences $(a_n)_{n \geq 0}$ and $(b_n)_{n \geq 0}$, defined by

$$a_{n+1} = \mathcal{H}(a_n, b_n) \text{ and } b_{n+1} = \mathcal{G}(a_{n+1}, b_n),\ n \geq 0, \tag{1.1}$$

for some initial arbitrarily chosen values a_0 and b_0. The main property of these sequences is given in the following theorem.

Theorem 1. *The sequences* $(a_n)_{n\geq 0}$ *and* $(b_n)_{n\geq 0}$*, defined by (1.1), are monotonously convergent to a common limit* $\mathcal{H} \boxtimes \mathcal{G}(a_0, b_0)$.

Proof. It is easy to verify the relations

$$\min(a,b) \leq \mathcal{G}(a,b) \leq \mathcal{A}(a,b) \leq \max(a,b), \tag{1.2}$$

famous under the name of arithmetic mean – geometric mean inequality. Their direct consequence is

$$\min(a,b) \leq \mathcal{H}(a,b) \leq \mathcal{G}(a,b) \leq \max(a,b). \tag{1.3}$$

Suppose that $0 < a_0 < b_0$. It follows that $a_0 < a_1 < b_0$, then $a_1 < b_1 < b_0$ and, generally, the assumption $a_{n-1} < b_{n-1}$ leads to $a_{n-1} < a_n < b_{n-1}$ and to $a_n < b_n < b_{n-1}$. Therefore, by mathematical induction, it was proven that

$$0 < a_0 < a_1 < \cdots < a_n < b_n < \cdots < b_1 < b_0.$$

The sequence $(a_n)_{n\geq 0}$ is thus monotonically increasing and bounded above by b_0. Therefore, it has a limit, say α. Similarly, the sequence $(b_n)_{n\geq 0}$ is monotonically decreasing and bounded below by a_0. So, it has the limit β. Passing to the limit in the relation $b_{n+1} = \mathcal{G}(a_{n+1}, b_n)$, we get $\beta = \mathcal{G}(\alpha, \beta)$, thus $\alpha = \beta$. Similar results, with the a's and b's interchanged, can be obtained in the case $0 < b_0 < a_0$. □

In the next two theorems, the value of the common limit of the sequences $(a_n)_{n\geq 0}$ and $(b_n)_{n\geq 0}$ is given. These were determined in Phillips (1981).

Theorem 2. *If* $0 < b_0 < a_0$*, the common limit of the sequences* $(a_n)_{n\geq 0}$ *and* $(b_n)_{n\geq 0}$ *is*

$$\mathcal{H} \boxtimes \mathcal{G}(a_0, b_0) = \frac{a_0 b_0}{\sqrt{a_0^2 - b_0^2}} \arccos \frac{b_0}{a_0}.$$

Proof. As in Archimedes' case, let

$$a_0 = \lambda \cdot \tan\theta \text{ and } b_0 = \lambda \cdot \sin\theta,$$

where $\lambda > 0$ and $0 < \theta < \frac{\pi}{2}$. Therefore,

$$\cos\theta = \frac{b_0}{a_0} \text{ and } \sin\theta = \frac{b_0}{\lambda},$$

thus

$$\theta = \arccos\frac{b_0}{a_0}.$$

The basic relation $\sin^2\theta + \cos^2\theta = 1$ gives

$$\lambda = \frac{a_0 b_0}{\sqrt{a_0^2 - b_0^2}}.$$

It is easy to see that

$$a_1 = 2\lambda \cdot \tan\frac{\theta}{2} \text{ and } b_1 = 2\lambda \cdot \sin\frac{\theta}{2},$$

and, generally, by an induction argument,

$$a_n = 2^n\lambda \cdot \tan\frac{\theta}{2^n} \text{ and } b_n = 2^n\lambda \cdot \sin\frac{\theta}{2^n}.$$

Of course

$$\lim_{n\to\infty} a_n = \lim_{n\to\infty} b_n = \lambda \cdot \theta,$$

which gives the desired result. □

Remark 1. Regarding the denotation $\mathcal{H} \boxtimes \mathcal{G}$, please refer to Chapter 3 Section 3.1 for general definitions and results.

Corollary 1. *In Archimedes' case, as*

$$a_0 = P_3 = 3\sqrt{3} \text{ and } b_0 = p_3 = 3\sqrt{3}/2,$$

the common limit is π.

Remark 2. To illustrate the resulting approximation process of π, we have the following table:

n	a_n	b_n
0	5.1961...	2.5980...
1	3.4641...	3.0000...
2	3.2153...	3.1058...
3	3.1596...	3.1326...
4	3.1460...	3.1393...
5	3.1427...	3.1410...
6	3.1418...	3.1414...

Theorem 3. *If* $0 < a_0 < b_0$, *the common limit of the sequences* $(a_n)_{n\geq 0}$ *and* $(b_n)_{n\geq 0}$ *is*

$$\mathcal{H} \boxtimes \mathcal{G}(a_0, b_0) = \frac{a_0 b_0}{\sqrt{b_0^2 - a_0^2}} \cosh^{-1}\left(\frac{b_0}{a_0}\right).$$

Proof. In this case, we can let

$$a_0 = \lambda \cdot \tanh\theta \text{ and } b_0 = \lambda \cdot \sinh\theta.$$

Therefore,

$$\cosh\theta = \frac{b_0}{a_0} \text{ and } \sinh\theta = \frac{b_0}{\lambda}$$

which gives

$$\theta = \cosh^{-1}\left(\frac{b_0}{a_0}\right)$$

and from the basic relation $\cosh^2\theta - \sinh^2\theta = 1$,

$$\lambda = \frac{a_0 b_0}{\sqrt{b_0^2 - a_0^2}}.$$

We have

$$a_1 = 2\lambda \cdot \tanh\frac{\theta}{2} \text{ and } b_1 = 2\lambda \cdot \sinh\frac{\theta}{2},$$

and, generally, by an induction argument,

$$a_n = 2^n\lambda \cdot \tanh\frac{\theta}{2^n} \text{ and } b_n = 2^n\lambda \cdot \sinh\frac{\theta}{2^n}.$$

Again,

$$\lim_{n\to\infty} a_n = \lim_{n\to\infty} b_n = \lambda \cdot \theta,$$

which gives the desired result. □

Corollary 2. *If we choose as initial values*

$$a_0 = 2t \text{ and } b_0 = t^2 + 1,\ t > 1,$$

the common limit of the sequences $(a_n)_{n\geq 0}$ and $(b_n)_{n\geq 0}$ is

$$\frac{2t(t^2+1)}{t^2-1}\log t.$$

Proof. We have

$$\cosh^{-1} x = \log(x + \sqrt{x^2 - 1}),\ x \geq 1,$$

so that for $x = b_0/a_0 = (t^2+1)/2t$, we obtain $\cosh^{-1} x = \log t$. □

Remark 3. If we choose $t = 2$, thus $a_0 = 4$ and $b_0 = 5$, we have the following table of values:

n	a_n	b_n
0	4.0000...	5.0000...
1	4.4444...	4.7140...
2	4.5752...	4.6441...
3	4.6094...	4.6267...
4	4.6180...	4.6224...
5	4.6202...	4.6213...
6	4.6208...	4.6210...

Multiplying by $3/20$, we get the following evaluation:

$$0.69312\ldots < \log 2 < 0.69316\ldots$$

As it is done in Phillips (1981) for the Archimedes' original algorithm, we can prove the following theorem:

Theorem 4. *The rate of convergence of the sequences* $(a_n)_{n\geq 0}$ *and* $(b_n)_{n\geq 0}$ *can be evaluated by the relation*

$$\lim_{n\to\infty} \frac{\mathcal{H} \boxtimes \mathcal{G}(a_0, b_0) - a_{n+1}}{\mathcal{H} \boxtimes \mathcal{G}(a_0, b_0) - a_n} = \lim_{n\to\infty} \frac{\mathcal{H} \boxtimes \mathcal{G}(a_0, b_0) - b_{n+1}}{\mathcal{H} \boxtimes \mathcal{G}(a_0, b_0) - b_n} = \frac{1}{4}, \ \forall a_0, b_0.$$

Proof. Assume $0 < b_0 < a_0$ and consider $b_n = 2^n \lambda \cdot \sin \frac{\theta}{2^n}$, as it was given above. Using MacLaurin's formula for the sinus function, we have

$$b_n = 2^n \lambda \cdot \left(\frac{\theta}{2^n} - \frac{\theta^3}{6 \cdot 2^{3n}} + \frac{\theta^4}{24 \cdot 2^{4n}} \cdot \sin \frac{\theta \cdot t_n}{2^n} \right), \ t_n \in (0, 1),$$

or

$$\lambda\theta - b_n = \lambda \cdot \left(\frac{\theta^3}{6 \cdot 2^{2n}} - \frac{\theta^4}{24 \cdot 2^{3n}} \cdot \sin \frac{\theta \cdot t_n}{2^n} \right).$$

Therefore,

$$\frac{\lambda\theta - b_{n+1}}{\lambda\theta - b_n} = \frac{\frac{1}{2^2} - \frac{\theta}{4 \cdot 2^{n+3}} \cdot \sin \frac{\theta \cdot t_{n+1}}{2^{n+1}}}{1 - \frac{\theta}{4 \cdot 2^n} \cdot \sin \frac{\theta \cdot t_n}{2^n}},$$

which gives the desired result in this case. The other three cases can be treated similarly. □

The result can be also given as follows:

Theorem 5. *The error of the sequences* $(a_n)_{n\geq 0}$ *and* $(b_n)_{n\geq 0}$ *tends to zero asymptotically like* $1/4^n$.

1.2 HERON'S METHOD OF EXTRACTING SQUARE ROOTS

Another ancient algorithm of the same nature is Heron's method of extracting square roots. As it is shown in Bullen (2003) (following Heath, 1949, Chajoth, 1932 and Pasche, 1946, 1948), Heron used the iteration of the arithmetic and harmonic means of two numbers to compute their geometric mean.

To find the square root of a positive number x, choose two numbers a, b with $0 < a < b$ and $ab = x$. Setting $a_o = a$ and $b_o = b$, define

$$a_{n+1} = \mathcal{H}(a_n, b_n),\ b_{n+1} = \mathcal{A}(a_n, b_n),\ n \geq 0.$$

We obtain the following result:

Theorem 6. *The sequences $(a_n)_{n\geq 0}$ and $(b_n)_{n\geq 0}$ are convergent to the common limit*

$$\mathcal{H} \otimes \mathcal{A}(a, b) = \mathcal{G}(a, b) = \sqrt{x}.$$

Proof. From (1.2) and (1.3) we have

$$\min(a, b) \leq \mathcal{H}(a, b) \leq \mathcal{A}(a, b) \leq \max(a, b).$$

It follows that

$$a_n < a_{n+1} < b_{n+1} < b_n\ ,\ n \geq 0.$$

Also, it is easy to see that

$$b_{n+1} - a_{n+1} = \frac{(b_n - a_n)^2}{2\,(b_n + a_n)} < \frac{b_n - a_n}{2},$$

thus

$$b_n - a_n < \frac{b - a}{2^n},\ n > 0.$$

Taking into account the relation

$$a_n b_n = a_{n-1} b_{n-1} = \ldots = a_o b_o = ab = x,$$

we deduce that the sequences $(a_n)_{n\geq 0}$ and $(b_n)_{n\geq 0}$ are convergent and

$$\lim_{n\to\infty} a_n = \lim_{n\to\infty} b_n = \sqrt{x} = \mathcal{G}(a, b).$$ □

Remark 4. Some comments on this iteration are given in Mathieu (1879) and Nowicki (1998). Regarding the notation $\mathcal{H} \otimes \mathcal{A}$, refer to Chapter 3 Section 3.5

for general definitions and results. We can illustrate Heron's approximation process by computing $\sqrt{2}$. Starting with $a=1, b=2$ we obtain the following table:

n	a_n	b_n
0	1.00000...	2.00000...
1	1.33333...	1.50000...
2	1.41176...	1.41666...
3	1.41420...	1.41421...

Remark 5. Heron's method has been extended to roots of higher order in Nikolaev (1925), Ory (1938) and Georgakis (2002). Also an iterative method for approximating higher order roots by using square roots has been given in Vythoulkas (1949).

1.3 LAGRANGE AND THE DEFINITION OF THE $\mathcal{AGM}$

It is generally accepted that the definition of the arithmetic–geometric mean ($\mathcal{AGM}$ for short) was given in Lagrange (1784–1785). Let us develop the ideas presented in Cox (1984), related to the above-mentioned paper.

Among other things, Lagrange was concerned with integrals of the form

$$\int \frac{N(y^2)dy}{\sqrt{(1+p^2y^2)(1+q^2y^2)}}, \tag{1.4}$$

where N is a rational function and $p>q>0$. Using the substitutions

$$p' = p + \sqrt{p^2 - q^2},$$
$$q' = p - \sqrt{p^2 - q^2},$$

and

$$y' = y\sqrt{\frac{1+p^2y^2}{1+q^2y^2}}, \tag{1.5}$$

he showed that

$$\frac{dy}{\sqrt{(1+p^2y^2)(1+q^2y^2)}} = \frac{dy'}{\sqrt{(1+p'^2y'^2)(1+q'^2y'^2)}}. \tag{1.6}$$

Using this, J.L. Lagrange intended to consider an iterative method. Taking

$$p_0 = p,\ q_0 = q$$

and

$$p_{n+1} = p_n + \sqrt{p_n^2 - q_n^2}\,,\; q_{n+1} = p_n - \sqrt{p_n^2 - q_n^2}\,,\; n \geq 0,$$

he noticed that the sequence $(p_n)_{n\geq 0}$ approaches $+\infty$ while $(q_n)_{n\geq 0}$ approaches 0. Thus the above substitution (1.5) cannot be used indefinitely in the integral (1.4), to arrive at an easily computable integral.

For this reason J.L. Lagrange considered a second iterative method. He remarked that the above substitutions can be reversed:

$$p = \frac{p' + q'}{2},\; q = \sqrt{p'q'},$$

and

$$y = \frac{\sqrt{2}}{p' + q'}\sqrt{p'q'y'^2 - 1 + \sqrt{(1 + p'^2y'^2)(1 + q'^2y'^2)}}.$$

Let us remark that

$$p = \mathcal{A}\left(p', q'\right),\; q = \mathcal{G}\left(p', q'\right).$$

Of course, the relation (1.6) remains true. Consider the integral

$$\int \frac{N'(y'^2)dy'}{\sqrt{(1 + p'^2y'^2)(1 + q'^2y'^2)}},$$

where N' is a rational function. Using the above substitutions, we obtain the integral

$$\int N'\left(\frac{y^2(1 + p^2y^2)}{1 + q^2y^2}\right)\frac{dy}{\sqrt{(1 + p^2y^2)(1 + q^2y^2)}} = \int \frac{N(y^2)dy}{\sqrt{(1 + p^2y^2)(1 + q^2y^2)}},$$

where N is again a rational function. The approximation method defined by Lagrange is based on the following double sequence: beginning with the terms

$$a_0 = p',\; b_0 = q',$$

define

$$a_{n+1} = \mathcal{A}(a_n, b_n),\; b_{n+1} = \mathcal{G}(a_n, b_n),\; n \geq 0. \tag{1.7}$$

Let us also denote in the following:

$$N_0 = N',\; y_0 = y',$$

$$y_{n+1} = \frac{\sqrt{2}}{a_n + b_n}\sqrt{a_nb_ny_n^2 - 1 + \sqrt{(1 + a_n^2y_n^2)(1 + b_n^2y_n^2)}}$$

and

$$N_{n+1}(y_{n+1}^2) = N_n\left(\frac{y_{n+1}^2(1+a_{n+1}^2 y_{n+1}^2)}{1+b_{n+1}^2 y_{n+1}^2}\right).$$

Taking into account the above formulas, the integral

$$\int \frac{N_0(y_0^2)dy_0}{\sqrt{(1+a_0^2y_0^2)(1+b_0^2y_0^2)}},$$

becomes, step by step,

$$\int \frac{N_n(y_n^2)dy_n}{\sqrt{(1+a_n^2y_n^2)(1+b_n^2y_n^2)}}, \quad n=1,2,... \tag{1.8}$$

On the other hand, using the relation between the means $\mathcal{A}$ and $\mathcal{G}$ we can prove the following results.

Theorem 7. *For each starting values a_0 and b_0, the sequences $(a_n)_{n\geq 1}$ and $(b_n)_{n\geq 1}$ defined by (1.7) have the following properties:*

$$a_1 \geq a_2 \geq \cdots \geq a_n \geq a_{n+1} \geq \cdots \geq b_{n+1} \geq b_n \geq \cdots \geq b_2 \geq b_1; \tag{1.9}$$

$$0 \leq a_n - b_n \leq \frac{|a_0 - b_0|}{2^n}. \tag{1.10}$$

Proof. Using (1.2) for $n > 0$, we have

$$a_n \geq \mathcal{A}(a_n, b_n) = a_{n+1} \geq b_{n+1} = \mathcal{G}(a_n, b_n) \geq b_n,$$

which results in (1.9). Concerning the second relation, we begin with

$$0 \leq a_1 - b_1 = \frac{a_0+b_0}{2} - \sqrt{a_0b_0} \leq \frac{a_0+b_0}{2} - \min(a_0,b_0) = \frac{|a_0-b_0|}{2}.$$

Then, from $b_{n+1} \geq b_n$ we obtain

$$a_{n+1} - b_{n+1} \leq a_{n+1} - b_n = \frac{a_n - b_n}{2},$$

which, by induction, results in (1.10). □

Remark 6. We can add to the relations (1.9) the inequalities

$$\min(a_0,b_0) \leq \mathcal{G}(a_0,b_0) = b_1 \leq a_1 = \mathcal{A}(a_0,b_0) \leq \max(a_0,b_0). \tag{1.11}$$

Corollary 3. *The sequences $(a_n)_{n\geq 1}$, $(b_n)_{n\geq 1}$ defined by (1.7) are convergent to a common limit $l = \mathcal{M}(a_0, b_0)$.*

Remark 7. From (1.9) and (1.11) we obtain

$$\min(a_0, b_0) \leq \mathcal{M}(a_0, b_0) \leq \max(a_0, b_0).$$

As we shall see later, this implies that $\mathcal{M}$ defines a mean, known as the **arithmetic–geometric mean** (or $\mathcal{AGM}$).

Corollary 4. *If $N' = 1$, the sequence of integrals (1.8) tends to the easily computable integral*

$$\int \frac{dy}{1 + l^2 y^2}.$$

Remark 8. Generally the convergence is much faster than it is suggested by (1.10). To illustrate this, let us show the evaluation of $\mathcal{M}(\sqrt{2}, 1)$ given in Gauss (1800). Using the following table:

n	a_n	b_n
0	1.414213562373095048802	1.000000000000000000000
1	1.207106781186547524401	1.189207115002721066717
2	1.198156948094634295559	1.198123521493120122607
3	1.198140234793877209083	1.198140234677307205798
4	1.198140234735592207441	1.198140234735592207439

Gauss found

$$\mathcal{M}(\sqrt{2}, 1) = 1.19814023473559220744... \tag{1.12}$$

Thus, in only four iterations, he obtained $\mathcal{M}$ to an accuracy of 19 places after the decimal point.

Remark 9. In Almkvist and Berndt (1988) another quantitative measure of the rapidity of convergence of the sequences $(a_n)_{n\geq 1}$ and $(b_n)_{n\geq 1}$ is given. Defining

$$c_n = \sqrt{a_n^2 - b_n^2}\,,\ n \geq 0,$$

we can see that

$$c_{n+1} = \frac{a_n - b_n}{2} = \frac{c_n^2}{4 \cdot a_{n+1}} < \frac{c_n^2}{4 \cdot M(a, b)}.$$

Thus $(c_n)_{n\geq 0}$ tends to 0 quadratically. Remember that in general, the convergence of the sequence $(\alpha_n)_{n\geq 0}$ to L is of the m-th order if the constants $C > 0$ and $m \geq 1$ exist such that

$$|\alpha_{n+1} - L| \leq C \cdot |\alpha_n - L|^m\ ,\ n \geq 0.$$

1.4 LEMNISCATIC INTEGRALS

Continuing with the history of the $\mathcal{AGM}$, let us again follow Cox (1984), this time focusing on the definition and study of the lemniscate.

In 1691, Jacob Bernoulli [Bernoulli (1744)] found the equation of the so-called **elastic curve**. This elastic curve, which resembles a parabola, is generated by a thin elastic rod which is bent until the two ends are perpendicular to the initial line position L. Assuming its vertex to be in the origin O and the axis of symmetry to be Ox, and further assuming that the distance of O to the line L is equal to 1, Jacob Bernoulli was able to show that the upper half of the curve is given by the equation

$$y = \int_0^x \frac{z^2 dz}{\sqrt{1 - z^4}}.$$

He also proved that the length of the rod is given by

$$\omega = 2 \int_0^1 \frac{dz}{\sqrt{1 - z^4}}. \tag{1.13}$$

Being well aware of the transcendental nature of the elastic curve, Jacob Bernoulli sought an algebraic curve with the same arc length. In 1694 he discovered **the lemniscate**, a curve which has the form of Figure 8 on its side and having the equation

$$x^2 + y^2 = a\sqrt{x^2 - y^2}.$$

For $a = 1$, he proved that one-half of the arc length of the lemniscate is just ω given by (1.13).

The elastic curve and the lemniscate appeared in many papers written in the eighteenth century. For example, in Stirling (1730) the following approximations are given:

$$\int_0^1 \frac{dz}{\sqrt{1 - z^4}} = 1.31102877714605987\ldots$$

and

$$\int_0^1 \frac{z^2 dz}{\sqrt{1 - z^4}} = 0.59907011736779611\ldots \tag{1.14}$$

Similar evaluations are given in Euler (1786), where the following interesting result is also proven:

$$\int_0^1 \frac{dz}{\sqrt{1 - z^4}} \cdot \int_0^1 \frac{z^2 dz}{\sqrt{1 - z^4}} = \frac{\pi}{4}. \tag{1.15}$$

These results, known to C.F. Gauss, were tied by him to the study of the $\mathcal{AGM}$. In the letter dated April 16, 1816, to his friend H. C. Schumacher (see Almkvist and Berndt, 1988), Gauss confided that he independently discovered the arithmetic–geometric mean in 1791 at the age of 14. He worked on the $\mathcal{AGM}$, the elastic curve, the lemniscate, and other related subjects for many years. Unfortunately, during his lifetime, Gauss only published one paper dealing with these subjects. It appeared in 1818 and it was in fact devoted to secular perturbations. Only with the publication of his collected works in Gauss (1876–1927), its contributions became apparent. Through this collection it is now known that in 1799 Gauss computed the value of $\mathcal{M}(\sqrt{2},1)$, as it is given in (1.12). Using (1.14) he remarked that

$$\mathcal{M}(\sqrt{2},1)=2\cdot\int_0^1\frac{z^2dz}{\sqrt{1-z^4}}$$

and from (1.15) he deduced that

$$\mathcal{M}(\sqrt{2},1)=\frac{\pi}{\omega},$$

its value being accurate up to eleventh place after the decimal point. Seeing the significance of this equality, in the 98th entry of his mathematical diary, Gauss noticed on May 30, 1799 that: "... the demonstration of this fact will surely open an entirely new field of analysis."

Gauss based this prophecy on his own findings. Some of his results will be presented here, while saving the most important ones for the next section. Motivated by the analogy to the circular functions, Gauss defined the lemniscatic functions by the relations:

$$\sin lemn\left(\int_0^x\frac{dz}{\sqrt{1-z^4}}\right)=x$$

and

$$\cos lemn\left(\frac{\omega}{2}-\int_0^x\frac{dz}{\sqrt{1-z^4}}\right)=x\,.$$

We stress that in this case ω plays the same role as does π for circular functions. Using the abbreviations sl and cl for the lemniscatic functions, Gauss proved a lot of formulas, such as

$$sl^2\phi+cl^2\phi+sl^2\phi\cdot cl^2\phi=1$$

and

$$sl\,(\phi+\psi)=\frac{sl\phi\cdot cl\psi+cl\phi\cdot sl\psi}{1-sl\phi\cdot sl\psi\cdot cl\phi\cdot cl\psi}\,.$$

He also defined these functions for complex numbers and proved that $sl\ \phi$ is double periodic, with periods 2ω and $2i\omega$.

We remark that Gauss also studied the $\mathcal{AGM}$ for complex numbers. Please refer to the paper Cox (1984) for more information on this subject.

1.5 ELLIPTIC INTEGRALS

Let us return to the double sequence $(a_n)_{n\geq 0}$ and $(b_n)_{n\geq 0}$, defined by J.L. Lagrange for the starting values

$$a_0 = a,\ b_0 = b$$

by the recurrences

$$a_{n+1} = \mathcal{A}(a_n, b_n),\ b_{n+1} = \mathcal{G}(a_n, b_n),\ n \geq 0.$$

As we have seen, they have the common limit $\mathcal{M}(a,b)$, which represents the $\mathcal{AGM}$ of a and b.

From the definition of $\mathcal{M}(a,b)$ we see that it has two obvious properties,

$$\mathcal{M}(a,b) = \mathcal{M}(a_1,b_1) = \mathcal{M}(a_2,b_2) = \ldots$$

and

$$\mathcal{M}(\lambda a, \lambda b) = \lambda \mathcal{M}(a,b).$$

In spite of this, the determination of $\mathcal{M}(a,b)$ is not at all a simple exercise. We have seen that C.F. Gauss calculated $\mathcal{M}(\sqrt{2},1)$ with high accuracy. In fact, he was able to prove much more.

Theorem 8. *If $a \geq b > 0$, then*

$$\mathcal{M}(a,b) = \frac{\pi}{2} \cdot \left[\int_0^{\pi/2} \frac{d\theta}{\sqrt{a^2\cos^2\theta + b^2\sin^2\theta}}\right]^{-1}. \tag{1.16}$$

Proof. Denote

$$I(a,b) = \int_0^{\pi/2} \frac{d\theta}{\sqrt{a^2\cos^2\theta + b^2\sin^2\theta}}. \tag{1.17}$$

The key step is to show that

$$I(a_1,b_1) = I(a,b), \tag{1.18}$$

thus

$$I(\mathcal{A}(a,b), \mathcal{G}(a,b)) = I(a,b). \tag{1.19}$$

To prove this, C.F. Gauss introduced a new variable θ' such that

$$\sin\theta = \frac{2a\sin\theta'}{a+b+(a-b)\sin^2\theta'} \tag{1.20}$$

and remarked that

$$(a^2\cos^2\theta + b^2\sin^2\theta)^{-1/2}d\theta = (a_1^2\cos^2\theta' + b_1^2\sin^2\theta')^{-1/2}d\theta'. \quad \square \tag{1.21}$$

The details of this proof are rather complicated, even as they were given in Jacobi (1881). Much simpler is the proof of D.J. Newman (Newman (1982)) given in Ganelius et al. (1982) and then in Schoenberg (1982). Changing the variable in (1.17) by the substitution

$$x = b\cdot\tan\theta,$$

we have

$$dx = b\cdot\frac{d\theta}{\cos^2\theta},$$

or

$$\frac{d\theta}{\cos\theta} = \frac{dx}{\sqrt{b^2+x^2}}.$$

This results in

$$I(a,b) = \int_0^{\pi/2}\frac{1}{\sqrt{a^2+b^2\tan^2\theta}}\cdot\frac{d\theta}{\cos\theta} = \int_0^{\infty}\frac{1}{\sqrt{a^2+x^2}}\cdot\frac{dx}{\sqrt{b^2+x^2}}.$$

Denoting

$$J(a,b) = \int_0^{\infty}\frac{dx}{\sqrt{(a^2+x^2)(b^2+x^2)}},$$

the following equality was obtained:

$$I(a,b) = J(a,b).$$

Therefore, for (1.18) we have to prove

$$J(a_1,b_1) = J(a,b). \tag{1.22}$$

We have

$$J(a_1,b_1) = \frac{1}{2}\int_{-\infty}^{\infty}\frac{dt}{\sqrt{(a_1^2+t^2)(b_1^2+t^2)}}$$

$$= \int_{-\infty}^{\infty} \frac{dt}{\sqrt{(a^2+2ab+b^2+4t^2)(ab+t^2)}}.$$

Changing the variable by the substitution

$$t = \frac{x^2 - ab}{2x},$$

we have

$$dt = \frac{1}{2}\left(1 + \frac{ab}{x^2}\right) dx,$$

thus

$$J(a_1, b_1) = \int_0^{\infty} \frac{\left(x^2 + ab\right) dx}{\sqrt{(a^2x^2 + b^2x^2 + a^2b^2 + x^4)(a^2b^2 + 2abx^2 + x^4)}} = J(a, b).$$

Iterating (1.18) gives us

$$I(a, b) = I(a_1, b_1) = I(a_2, b_2) = \cdots \tag{1.23}$$

therefore

$$I(a, b) = \lim_{n\to\infty} I(a_n, b_n) = I(l, l) = \frac{\pi}{2 \cdot l}, \tag{1.24}$$

where

$$l = \lim_{n\to\infty} a_n = \lim_{n\to\infty} b_n = \mathcal{M}(a, b).$$

Thus

$$\mathcal{M}(a, b) = \frac{\pi}{2 \cdot I(a, b)},$$

which is (1.16).

Remark 10. Writing (1.19) as

$$f(\mathcal{A}(a, b), \mathcal{G}(a, b)) = f(a, b) \tag{1.25}$$

this is named **Gauss' functional equation**. As in (1.23) and (1.24), its solution gives a representation of $\mathcal{M}$. For instance, iterating (1.22), we get the second representation

$$\mathcal{M}(a, b) = \frac{\pi}{2 \cdot J(a, b)}.$$

Remark 11. As it is shown in Cox (1984), setting

$$py = \tan\theta,$$

one obtains

$$\frac{dy}{\sqrt{(1+p^2y^2)(1+q^2y^2)}} = \frac{d\theta}{\sqrt{p^2\cos^2\theta + q^2\sin^2\theta}},$$

so that the relation (1.6) results in (1.21). Thus not only could Lagrange define the $\mathcal{AGM}$, he also could prove the above theorem effortlessly. D.A. Cox was of the opinion that "unfortunately, none of this happened; Lagrange never realized the power of what he had discovered".

Remark 12. Another representation of the $\mathcal{AGM}$ is given by a **complete elliptic integral of the first kind**, i.e. integrals of the form

$$K(x) = \int_0^{\pi/2} \frac{d\theta}{\sqrt{1-x^2\sin^2\theta}} = \int_0^1 \frac{dz}{\sqrt{(1-z^2)(1-x^2z^2)}}. \tag{1.26}$$

Indeed, we have

$$I(a,b) = \frac{1}{a}K(x), \text{ where } x = \frac{\sqrt{a^2-b^2}}{a},$$

thus

$$\mathcal{M}(a,b) = \frac{a\pi}{2\cdot K\left(\sqrt{1-\left(\frac{b}{a}\right)^2}\right)}. \tag{1.27}$$

See also Hofsommer and van de Riet (1962), van de Riet (1963), Salamin (1976), Muntean and Vornicescu (1993–1994), and Bullen (2003).

1.6 HYPERGEOMETRIC SERIES

The relation (1.27) is given also in the following simpler form in Almkvist and Berndt (1988).

Theorem 9. *If* $|x| < 1$ *then*

$$\mathcal{M}(1+x, 1-x) = \frac{\pi}{2\cdot K(x)}. \tag{1.28}$$

Proof. For

$$x = \frac{\sqrt{a^2-b^2}}{a}$$

we have

$$\begin{aligned}\mathcal{M}(1+x,1-x) &= \mathcal{M}(\mathcal{A}(1+x,1-x),\,\mathcal{G}(1+x,1-x))\\ &= \mathcal{M}\left(1,\sqrt{1-x^2}\right) = \mathcal{M}\left(1,\frac{b}{a}\right) = \frac{1}{a}\mathcal{M}(a,b),\end{aligned}$$

thus (1.28). □

On the other hand, using the binomial series

$$(1+x)^\alpha = \sum_{n=0}^{\infty} \frac{(\alpha)_n}{n!}\cdot x^n, \tag{1.29}$$

where

$$(\alpha)_n = \alpha\cdot(\alpha-1)\cdot\ldots\cdot(\alpha-n+1),\ \alpha\in\mathbb{R}, \tag{1.30}$$

we have

$$(1-t)^{-\frac{1}{2}} = \sum_{n=0}^{\infty}\left(\frac{1}{2}\right)_n \frac{t^n}{n!},$$

thus

$$K(x) = \int_0^{\frac{\pi}{2}} \sum_{n=0}^{\infty}\left(\frac{1}{2}\right)_n \frac{(x\sin\theta)^{2n}}{n!} d\theta.$$

Integrating termwise, we find that

$$K(x) = \sum_{n=0}^{\infty}\left(\frac{1}{2}\right)_n \frac{x^{2n}}{n!}\int_0^{\frac{\pi}{2}} \sin^{2n}\theta\, d\theta.$$

For the integral

$$I_{2n} = \int_0^{\frac{\pi}{2}} \sin^{2n}\theta\, d\theta,$$

using the transformation

$$I_{2n+2} = I_{2n} - \int_0^{\frac{\pi}{2}} \cos\theta\left(\frac{\sin^{2n+1}\theta}{2n+1}\right)' d\theta,$$

we get the recurrence

$$I_{2n+2} = \frac{2n+1}{2n+2}\cdot I_{2n},$$

thus the well-known expression (see Bierens de Haan, 1867)

$$I_{2n} = \left(\frac{1}{2}\right)_n \cdot \frac{1}{n!} \cdot \frac{\pi}{2}.$$

Finally we have

$$K(x) = \frac{\pi}{2} \sum_{n=0}^{\infty} \left(\frac{1}{2}\right)_n^2 \frac{x^{2n}}{(n!)^2}.$$

Considering **Gauss' ordinary hypergeometric series**

$$F(a, b, c; x) = \sum_{n=0}^{\infty} \frac{(a)_n \cdot (b)_n}{(c)_n \cdot n!} \cdot x^n, \ |x| < 1,$$

we get

$$K(x) = \frac{\pi}{2} \cdot F\left(\frac{1}{2}, \frac{1}{2}, 1; x^2\right). \tag{1.31}$$

Thus, we deduce another representation of the $\mathcal{AGM}$.

Theorem 10. *If* $|x| < 1$ *then*

$$\mathcal{M}(1 + x, 1 - x) = \frac{1}{F\left(\frac{1}{2}, \frac{1}{2}, 1; x^2\right)}.$$

1.7 LANDEN'S TRANSFORMATION

In the proof of Theorem 8, the transformation (1.20) is written alternatively as

$$\sin\theta = \frac{(1 + k) \cdot \sin\theta'}{1 + k \cdot \sin^2\theta'},$$

which is commonly referred to as **the Gauss transformation**. In a similar way, for a direct proof of Theorem 9 the transformation defined in Landen (1771) can be used by letting

$$\tan\theta_1 = \frac{\sin(2\theta)}{x_1 + \cos(2\theta)},$$

where

$$x_n = \sqrt{1 - \left(\frac{b_n}{a_n}\right)^2}.$$

This is called **Landen's transformation** and it was used in Legendre (1825) in the following iterative procedure. It is rather difficult to prove that

$$K(x) = (1 + x_1) \cdot K(x_1),$$

but, upon n iterations, we have

$$K(x) = (1 + x_1) \cdot (1 + x_2) \cdot \ldots \cdot (1 + x_n) \cdot K(x_n).$$

Since

$$1 + x_k = \frac{1}{a_k} \left(\frac{a_{k-1} + b_{k-1}}{2} + \sqrt{\left(\frac{a_{k-1} + b_{k-1}}{2} \right)^2 - a_{k-1} \cdot b_{k-1}} \right) = \frac{a_{k-1}}{a_k},$$

we get

$$K(x) = \frac{a}{a_n} \cdot K(x_n).$$

By letting n to tend to ∞, we conclude that

$$K(x) = \frac{a}{M(a,b)} \cdot K(0) = \frac{a \cdot \pi}{2 \cdot M(a,b)},$$

which gives (1.28).

The following Landen's transformation for incomplete elliptic integrals of the first kind is also presented in Almkvist and Berndt (1988): if

$$x \sin \alpha = \sin(2\beta - \alpha),$$

then

$$(1 + x) \int_0^\alpha \left(1 - x^2 \sin^2 \theta\right)^{-1/2} d\theta = 2 \int_0^\beta \left(1 - \frac{4x}{(1+x)^2} \sin^2 \theta\right)^{-1/2} d\theta.$$

In the special case where

$$\alpha = \pi \ , \ \beta = \pi/2 \ , \ 0 \le x < 1,$$

it becomes

$$K(x) = \frac{1}{1+x} \cdot K\left(\frac{2\sqrt{x}}{1+x} \right). \tag{1.32}$$

In Borwein and Borwein (1987) this is called the "upward functional relation" satisfied by the complete elliptic integral. It is accomplished by the following "downward functional relation"

$$K(x) = \frac{2}{1+x'} \cdot K\left(\frac{1-x'}{1+x'}\right), \tag{1.33}$$

where $x' = \sqrt{1-x^2}$. Indeed, let us denote

$$y = \frac{1-x'}{1+x'}$$

and write (1.32) for y:

$$K\left(\frac{2\sqrt{y}}{1+y}\right) = (1+y) \cdot K(y).$$

As

$$x' = \frac{1-y}{1+y},$$

thus

$$x = \sqrt{1-x'^2} = \frac{2\sqrt{y}}{1+y},$$

we obtain (1.33).

In the same paper, the Landen's transformation for hypergeometric series is remembered:

$$F\left(a, b; 2b; \frac{4x}{(1+x)^2}\right) = (1+x)^{2a} \cdot F\left(a, a-b+\frac{1}{2}; b+\frac{1}{2}; x^2\right).$$

The special case

$$F\left(\frac{1}{2}, \frac{1}{2}; 1; \frac{4x}{(1+x)^2}\right) = (1+x) \cdot F\left(\frac{1}{2}, \frac{1}{2}; 1; x^2\right),$$

is also a consequence of the relations (1.31) and (1.32).

1.8 THE PERIMETER OF AN ELLIPSE

If an ellipse is given by the parametric equations

$$x = a\cos\theta, \ \ y = b\sin\theta, \ \ 0 \le \theta \le 2\pi,$$

then its perimeter is expressed by

$$L(a,b) = 4 \cdot \int_0^{\frac{\pi}{2}} \sqrt{a^2\sin^2\theta + b^2\cos^2\theta}\, d\theta.$$

Considering the **complete elliptic integral of the second kind**

$$E(x) = \int_0^{\pi/2} \sqrt{1 - x^2 \sin^2 \theta}\, d\theta, \ |x| < 1, \tag{1.34}$$

then

$$L(a,b) = 4a \cdot E(\varepsilon)$$

where

$$\varepsilon = \frac{\sqrt{a^2 - b^2}}{a}$$

denotes the **eccentricity** of the ellipse.

The elliptic integrals $K(x)$ and $E(x)$ are related by many formulas. We mention here only the following formula given in Legendre (1825). If $0 < x < 1$ and $x' = \sqrt{1 - x^2}$, then

$$K(x)E(x') + K(x')E(x) - K(x)K(x') = \frac{\pi}{2}.$$

A simple proof of this formula may be found in Almkvist and Berndt (1988). Taking it into account, we deduce that as in the case of the $\mathcal{AGM}$, the value of $L(a,b)$ is very difficult to determine. The first approximations for $L(a,b)$ were given in 1609 by Kepler (see Kepler, 1860). Certainly, he needed them in astronomy. Using some non-rigorous arguments, he has given the lower bound

$$L(a,b) \approx 2\pi\sqrt{ab}$$

and the upper bound

$$L(a,b) \approx \pi(a+b).$$

In Almkvist and Berndt (1988) a table with 13 approximations for $L(a,b)$ is given. Generally they are derived from some exact formulas, which are of more interest for us because of their similarity to formulas related to the $\mathcal{AGM}$.

The first result was given in MacLaurin (1742).

Theorem 11. *If ε is the eccentricity of the ellipse, then*

$$L(a,b) = 2\pi a \cdot F\left(\frac{1}{2}, -\frac{1}{2}; 1; \varepsilon^2\right).$$

Proof. As in the proof of Theorem 9, we have

$$L(a,b) = 4a \int_0^{\pi/2} \sqrt{1 - \varepsilon^2 \cos^2 \theta}\, d\theta$$

$$= 4a \sum_{n=0}^{\infty} \left(-\frac{1}{2}\right)_n \frac{\varepsilon^{2n}}{n!} \int_0^{\pi/2} \cos^{2n}\theta \, d\theta$$

$$= 2\pi a \sum_{n=0}^{\infty} \left(-\frac{1}{2}\right)_n \left(\frac{1}{2}\right)_n \frac{\varepsilon^{2n}}{(n!)^2},$$

which gives the result. □

A second result was given in Euler (1774). As it is stated in Almkvist and Berndt (1988), this second result can be derived from MacLaurin's result via a certain quadratic transformation for hypergeometric series that is different from Landen's.

Theorem 12. *For every $a, b > 0$, the formula*

$$L(a,b) = \pi\sqrt{2\left(a^2+b^2\right)} \cdot F\left(-\frac{1}{4}, -\frac{1}{4}; 1; \left(\frac{a^2-b^2}{a^2+b^2}\right)^2\right),$$

is valid.

A third result was given in Ivory (1796).

Theorem 13. *If*

$$\lambda = \frac{a-b}{a+b},$$

then

$$L(a,b) = \pi\,(a+b) \cdot F\left(-\frac{1}{2}, -\frac{1}{2}; 1; \lambda^2\right).$$

Proof. If in Landen's transformation for hypergeometric series we set

$$a = -1/2,\ b = 1/2,\ x = \lambda,$$

we find that

$$F\left(-\frac{1}{2}, \frac{1}{2}; 1; e^2\right) = \frac{a+b}{2a} \cdot F\left(-\frac{1}{2}, -\frac{1}{2}; 1; \lambda^2\right).$$

Therefore, the result follows from MacLaurin's theorem. □

The link between the arithmetic–geometric mean and the length of an arc of an ellipse was also studied in Almkvist (1978).

Chapter 2

Means

ABSTRACT

In this chapter, we present the ten means known by the ancient Greeks, along with some of their properties and relations. We mention some methods for constructing bivariate means, and some of their properties and related results. Also, we present the notion of complementariness with respect to a mean, some properties, and we try to compute the complementary with respect to some families of means.

In presenting some classical results related to the arithmetic–geometric mean, we have used only three basic means: the arithmetic mean, the geometric mean, and the harmonic mean. These means are among the ten means known by the ancient Greeks. In this chapter, all of these means will be presented, along with some of their properties and relations. They will serve as good examples for the general theory that will follow.

We shall present a few notions and results related to means of two variables. The motivation behind this – besides the fact that these means will be used in the following chapters – was driven by a remark from Borwein and Borwein (1987) where it is stated that "there is a great literature on particular means and very little on means in general." This gap was filled in part by the books Bullen et al. (1988) and Bullen (2003), but they are devoted to means of several variables and, as a result, many special problems for means of two variables were omitted.

2.1 MEANS AND PROPERTIES OF MEANS

2.1.1 Greek means

As many other important Greek mathematical contributions, the means are presented by Pappus of Alexandria in his books, in the fourth century AD (see Pappus, 1932). Some indications about them can be found in the books Gini (1958), Borwein and Borwein (1987), and Antoine (1998). We present here a variant of the original construction of the means, but also their modern transcriptions. We selected some properties of these means and some relations among them, as they are given in Toader and Toader (2002b) and Toader and Toader (2005).

Pythagoras of Samos in the sixth century BC was already familiar with the arithmetic mean, the geometric mean, and the harmonic mean. In order to con-

Means in Mathematical Analysis. http://dx.doi.org/10.1016/B978-0-12-811080-5.00002-5

struct them he used the method of proportions. More specifically, he considered a set of three numbers with the property that two of their differences are in the same rapport as two of the initial numbers. We present the method in a way that leads more directly to our usual definitions.

Let $a > m > b > 0$. Then m represents:

1. the **arithmetic mean** of a and b if

$$\frac{a-m}{m-b} = \frac{a}{a};$$

2. the **geometric mean** of a and b if

$$\frac{a-m}{m-b} = \frac{a}{m} = \frac{m}{b};$$

3. the **harmonic mean** of a and b if

$$\frac{a-m}{m-b} = \frac{a}{b};$$

According to Antoine (1998), three other means were defined by Eudoxus and finally other four means by Temnoides and Euphranor. In Gini (1958), all these seven means are attributed to Nicomah. Of these seven means, only three have names:

4. the **contraharmonic mean** of a and b represented by

$$\frac{a-m}{m-b} = \frac{b}{a};$$

5. the **contrageometric mean** of a and b given by

$$\frac{a-m}{m-b} = \frac{b}{m};$$

6. the **second contrageometric mean** of a and b given by

$$\frac{a-m}{m-b} = \frac{m}{a}.$$

The rest of four not-named means are defined by the relations:

7.

$$\frac{a-m}{a-b} = \frac{b}{a};$$

8.

$$\frac{a-m}{a-b} = \frac{m}{a};$$

9.

$$\frac{a-b}{m-b}=\frac{a}{b};$$

10.

$$\frac{a-b}{m-b}=\frac{m}{b}.$$

These ten means are the only means which can be defined using the method of proportions, attributed to Pythagoras of Samos (569–500 BC) (see Gini, 1958; Antoine, 1998), but also to Eudoxus (see Eymard and Lafon, 2004). Having no access to original sources, we must content ourselves to present such controversies, without taking any adherent position.

Solving each of the above relations as an equation with the unknown term m, we obtain the analytic expressions $M(a, b)$ of the means. For the first four means we use the classical denotations. For the other six, we use the neutral symbols proposed in Borwein and Borwein (1987). As a result, in order, the following means are obtained:

1.

$$\mathcal{A}(a,b)=\frac{a+b}{2};$$

2.

$$\mathcal{G}(a,b)=\sqrt{ab};$$

3.

$$\mathcal{H}(a,b)=\frac{2ab}{a+b};$$

4.

$$\mathcal{C}(a,b)=\frac{a^2+b^2}{a+b};$$

5.

$$\mathcal{F}_5(a,b)=\frac{a-b+\sqrt{(a-b)^2+4b^2}}{2};$$

6.

$$\mathcal{F}_6(a,b)=\frac{b-a+\sqrt{(a-b)^2+4a^2}}{2};$$

7.

$$\mathcal{F}_7(a,b) = \frac{a^2 - ab + b^2}{a};$$

8.

$$\mathcal{F}_8(a,b) = \frac{a^2}{2a - b};$$

9.

$$\mathcal{F}_9(a,b) = \frac{b(2a - b)}{a};$$

10.

$$\mathcal{F}_{10}(a,b) = \frac{b + \sqrt{b(4a - 3b)}}{2}.$$

Occasionally, it is convenient to also denote

$$\mathcal{F}_1 = \mathcal{A},\ \mathcal{F}_2 = \mathcal{G},\ \mathcal{F}_3 = \mathcal{H} \text{ and } \mathcal{F}_4 = \mathcal{C}.$$

The first four expressions of the Greek means are symmetric, that is we can use them also for the definition of the corresponding means for $a < b$. For the other six expressions, we have to replace a with b to define the means in the case where $a < b$.

2.1.2 Definition and properties of means

There are more definitions of means as we can see in the book Gini (1958), or in papers such as Chisini (1929), Jecklin (1948, 1949), or Aczél et al. (1987). The most commonly used definition may be found in the book Hardy et al. (1934); this definition was even suggested by Cauchy (as it is stated in Gini, 1958).

Definition 1. A **mean** (on the interval J) is defined as a function $M : J^2 \to J$, which has the property

$$a \wedge b \leq M(a,b) \leq a \vee b,\ \forall a,b \in J,$$

where

$$a \wedge b = \min(a,b) \text{ and } a \vee b = \max(a,b).$$

Regarding the properties of means, of course, each mean is **reflexive**, that is

$$M(a,a) = a,\ \forall a \in J,$$

which could also be used as a definition of $M(a,a)$ in case it is necessary. Occasionally, this weaker condition is taken as a definition of means (see Aczél et al., 1987). Following Matkowski (2006), we prefer to use the name of pre-means for reflexive functions (though in many previous papers we called them generalized means).

Definition 2. A **pre-mean** (on the interval J) is defined as a reflexive function $M : J^2 \to J$.

A (pre-)mean can have additional properties.

Definition 3. The (pre-)mean M is called:

a) **symmetric**, if

$$M(a,b) = M(b,a),\ \forall a,b \in J;$$

b) **homogeneous** (of degree one), if

$$M(ta,tb) = t \cdot M(a,b),\ \forall t > 0\,, a,b \in J; \tag{2.1}$$

c) **(strictly) isotone**, if for $a,b \in J$ the functions $M(a,.)$ and $M(.,b)$ are (strictly) increasing;

d) **strict at the left**, if

$$M(a,b) = a \text{ implies } a = b,$$

strict at the right, if

$$M(a,b) = b \text{ implies } a = b,$$

and **strict**, if it is strict at the left and strict at the right.

Remark 13. In most cases we have $J = [0,\infty)$ or $J = (0,\infty)$, but there are also means with smaller domain of definition. For instance, some trigonometric means have $J = [0,\pi/2]$ or $J = [0,\pi/2)$. Take for example

$$M(a,b) = \arcsin\left(\frac{\sin a + \sin b}{2}\right), \tag{2.2}$$

which can be found in Gini (1958). If the mean is homogeneous, of course $J \supset (0,\infty)$. To avoid this restriction we could consider the relation (2.1) only in the neighborhood of 1.

In what follows, we shall use the following obvious lemma:

Lemma 1. *A mean M is isotone if and only if*

$$M(a,b) \le M(a',b'), \text{ for all } a \le a',\ b \le b'. \tag{2.3}$$

Example 1. We shall consider $\wedge$ and $\vee$ also as means defined by

$$\wedge(a,b) = a \wedge b, \vee(a,b) = a \vee b, \forall a,b > 0.$$

They are symmetric, homogeneous, and isotone, but are not strict neither at the left nor at the right.

Remark 14. In Borwein and Borwein (1987), these means are used for the definition of the Greek means. Namely, a is replaced by $a \vee b$ and b by $a \wedge b$. So we get expressions of the following type:

$$M(a \vee b, a \wedge b).$$

With this construction any mean becomes symmetric.

Remark 15. All the Greek means are homogeneous and strict. In Toader and Toader (2002b), the monotony of the above means is also studied (see also Toader and Toader, 2005). We have the following results.

Theorem 14. *For $a > b > 0$, the Greek means have the following monotonicities: 1) All the means are increasing with respect to a on (b,∞). 2) The means $\mathcal{A}, \mathcal{G}, \mathcal{H}, \mathcal{F}_6, \mathcal{F}_8, \mathcal{F}_9$ and $\mathcal{F}_{10}$ are increasing with respect to b on $(0,a)$, thus they are isotone. 3) For each of the means $\mathcal{C}$, $\mathcal{F}_5$, and $\mathcal{F}_7$ there is a number $0 < p < 1$ such that the mean is decreasing with respect to b on the interval $(0, p \cdot a)$ and increasing on $(p \cdot a, a)$. These means have the values $M(a,0) = M(a,a) = a$, respectively $M(a, pa) = qa$. The values of the constants p, q are given in the following table:*

M	$\mathcal{C}$	$\mathcal{F}_5$	$\mathcal{F}_7$
p	$\sqrt{2}-1$	$2/5$	$1/2$
q	$2(\sqrt{2}-1)$	$4/5$	$3/4$

Remark 16. Simple examples of non-symmetric means may be given by the projections Π_1 and Π_2 defined respectively by

$$\Pi_1(a,b) = a,\ \Pi_2(a,b) = b,\ \forall a,b \in J. \tag{2.4}$$

Now Π_1 is strict at the right and Π_2 is strict at the left. We can consider them as trivial Greek means defined by

$$\frac{a-b}{m-b} = \frac{a}{m},$$

respectively

$$\frac{a-b}{a-m}=\frac{b}{m}.$$

2.1.3 Quasi-arithmetic means

Important examples of means are given by the following definition, that can be found for example in Daróczy (2005). To present it, let us denote by $CM(J)$ the set of all strictly monotonous continuous functions $f : J \to \mathbb{R}$.

Definition 4. A mean M is called **quasi-arithmetic** if there is a function $f \in CM(J)$ such that

$$M(a,b)=f^{-1}\left(\frac{f(a)+f(b)}{2}\right),\ \forall a,b\in J.$$

We shall use the denotation $M = \mathcal{A}(f)$.

The most well-known example of quasi-arithmetic mean is given by the **power** (or **Hölder**) **mean** $\mathcal{P}_n = \mathcal{A}(e_n)$ where

$$e_n(x)=\begin{cases} x^n,\ n\neq 0\\ \ln x,\ n=0.\end{cases} \tag{2.5}$$

It follows that

$$\mathcal{P}_n(a,b)=\left(\frac{a^n+b^n}{2}\right)^{1/n}$$

for $n \neq 0$, while $\mathcal{P}_0 = \mathcal{G}$. We have $\mathcal{P}_{-1} = \mathcal{H}$, $\mathcal{P}_1 = \mathcal{A}$ and for $n = 2$ we denote $\mathcal{P}_2 = \mathcal{Q}$ the root-square mean.

In Aczél (1948), the following characterization of quasi-arithmetic means is provided. It is a continuation of the theorem of Kolmogorov and Nagumo from 1930, which characterizes the quasi-arithmetic means of any number of variables.

Theorem 15. *A mean M is quasi-arithmetic on a closed interval J if and only if it has the following properties: (i) continuity; (ii) strict isotony; (iii) reflexivity; (iv) symmetry; (v) bisymmetry, i.e.*

$$M(M(a,b),M(c,d))=M(M(a,c),M(b,d)),\ \forall a,b,c,d\in J.$$

More generally, given a mean M (on J) and a strictly monotonous function $f : J' \to J$, it is easy to see that the function $M(f)$ defined by

$$M(f)(a,b)=f^{-1}(M(f(a),f(b)))$$

is also a mean (on J'). By analogy with the quasi-arithmetic mean, $M(f)$ can be called **quasi-M mean**. This method was used for constructing other means, for example in Sándor et al. (1996) and Sándor and Toader (2002).

In Borwein and Borwein (1987), the following property is given.

Proposition 1. *The mean $M(f)$ is isotone, symmetric, or strict whenever M is isotone, symmetric, or strict. Moreover, $M(e_p)$ is homogeneous for $p \neq 0$, whenever M is.*

It is known (see Hardy et al., 1934) that

$$\lim_{n \to 0} \mathcal{P}_n = \mathcal{P}_0, \tag{2.6}$$

where we used the ordinary denotation $\lim_{n \to \infty} M_n = M$ for

$$\lim_{n \to \infty} M_n(a,b) = M(a,b), \ \forall a, b \in J.$$

The relation (2.6) is not very natural taking into account the expression of the function e_n. To make this idea clearer we shall consider another example, given in Toader (2000). Let the quasi-$\mathcal{H}$ mean $\mathcal{H}(e_n)$ be given by

$$\mathcal{H}(e_n)(a,b) = \left(\frac{2a^n b^n}{a^n + b^n} \right)^{1/n}, \ n \neq 0.$$

We have again that

$$\lim_{n \to 0} \mathcal{H}(e_n) = \mathcal{G},$$

but it is different from $\mathcal{H}(\ln)$ given by

$$\mathcal{H}(\ln)(a,b) = \exp\left(\frac{2 \cdot \ln a \cdot \ln b}{\ln a + \ln b} \right).$$

This can be explained as follows. Let us consider another family of functions $g_n : (1, \infty) \to (0, \infty)$ given by

$$g_n(x) = \begin{cases} \frac{x^n - 1}{n}, \ n \neq 0 \\ \ln x \ , \ n = 0. \end{cases}$$

We have now $\lim_{n \to 0} g_n = g_0$. As for $n \neq 0$, $\mathcal{A}(g_n) = \mathcal{A}(e_n) = \mathcal{P}_n$ (on $(1, \infty)$), naturally (2.6) holds. But

$$\mathcal{H}(g_n)(a,b) = \left(\frac{2a^n b^n - a^n - b^n}{a^n + b^n - 2} \right)^{1/n},$$

thus $\mathcal{H}(g_n) \neq \mathcal{H}(e_n)$. In this case we have

$$\lim_{n \to 0} \mathcal{H}(g_n) = \mathcal{H}(g_0).$$

In Gini (1958), additional examples of means can be found. Although some of these are defined using a modified method of proportions, they are quasi-arithmetic means $\mathcal{A}(f)$ with f defined respectively by

$$f(x) = c^x,\ f(x) = c^{\frac{1}{x}},\ f(x) = x^x,\ f(x) = x^{\frac{1}{x}}.$$

Additionally, for the mean (2.2) we have $M = \mathcal{A}(\sin)$. The means $\mathcal{A}(\cos)$, $\mathcal{A}(\tan)$, and $\mathcal{A}(\cot)$ are also considered.

A special case of the quasi-M mean considered in Borwein and Borwein (1987) is used to define another mean.

Lemma 2. *If M is an isotone, homogeneous mean, then for every $p \in \mathbb{R}$,*

$$_pM = \frac{[M(e_p)]^p}{[M(e_{p-1})]^{p-1}}$$

is a homogeneous mean, which is strict or symmetric whenever M is strictly isotone or symmetric.

The most well-known example of mean constructed this way is the p-th **contraharmonic mean** (or **Lehmer mean**)

$$\mathcal{C}_p(a,b) = {}_p\mathcal{A}(a,b) = \frac{a^p + b^p}{a^{p-1} + b^{p-1}},$$

which is a generalization of the contraharmonic mean $\mathcal{C} = \mathcal{C}_2$. It is also a special case of **Beckenbach–Gini means** defined by

$$\mathcal{C}_f(a,b) = \frac{af(a) + bf(b)}{f(a) + f(b)}, \forall a,b > 0,$$

where f is a positive function (see Bullen (2003, page 406)). In its turn, the Lehmer mean can be used for the building of a more general class of means – the **Gini means** (called in Bullen (1998) the **Gini–Dresher means**) or **sum means**:

$$\mathcal{S}_{r,s} = \mathcal{C}_p(e_{r-s}),\ p = \frac{r}{r-s},$$

which gives

$$\mathcal{S}_{r,s}(a,b) = \left(\frac{a^r + b^r}{a^s + b^s}\right)^{\frac{1}{r-s}},\ r \neq s,$$

and

$$\mathcal{S}_{r,r}(a,b) = \lim_{s \to r} \mathcal{S}_{r,s}(a,b) = \left(a^{a^r} \cdot b^{b^r}\right)^{\frac{1}{a^2+b^2}}.$$

We shall use also the special case of sum means $\mathcal{S}_p = \mathcal{S}_{p-1,1}$ (**special Gini mean**). Of course $\mathcal{S}_{p,p-1} = \mathcal{C}_p$, $\mathcal{S}_{0,0} = \mathcal{G}$, and denote $\mathcal{S}_{1,1} = \mathcal{J}$. Another special case is

$$\mathcal{S}_{r,r-1}(e_{-1}),$$

called **Moskovitz mean**.

Another generalization of the power means is given by the family of **Muirhead means** $\mathcal{V}_{p,q}$ defined for $p+q \neq 0$ by

$$\mathcal{V}_{p,q}(a,b) = \left(\frac{a^p b^q + a^q b^p}{2}\right)^{\frac{1}{p+q}}$$

(see Bullen, 2003 and Trif, 2006). We shall refer to the special case $\mathcal{V}_{p,1-p}$, also called **Heinz mean** (see Bhatia, 2006).

Let us remark that almost all of the above means are of the form

$$N_{r,s}(a,b) = \left[\frac{g_s(1,1)}{f_r(1,1)} \cdot \frac{f_r(a,b)}{g_s(a,b)}\right]^{\frac{1}{r-s}},$$

where f_r and g_s are homogeneous functions of degree r, respectively s.

Another type of means was defined in Matkowski, 2011. Let the real functions f and g be continuous, positive, and strictly monotonic in a real interval J such that $\frac{f}{g}$ is one-to-one. Then the function $Q^{[f,g]} : J^2 \to \mathbb{R}$ defined by

$$Q^{[f,g]}(a,b) := \left(\frac{f}{g}\right)^{-1} \left(\frac{f(a)}{g(b)}\right), \quad a,b \in J$$

is a strict mean in J called a **quotient mean**. We can remark that this happens if and only if f and g are of different type of strict monotonicity. The mean $Q^{[f,g]}$ is symmetric if and only if the product $f \cdot g$ is a constant function and then

$$Q^{[f,g]}(a,b) = f^{-1}(\sqrt{f(a) \cdot f(b)}) = g^{-1}(\sqrt{g(a) \cdot g(b)}), \quad a,b \in J.$$

2.1.4 Other methods for the construction of means

There are other additional general methods for the construction of means. Some of these methods use algebraic or functional relations like those from Moskovitz (1933), Stolarsky (1975, 1980), or Bauer (1986). Others are based on the mean value theorem for derivatives or for integrals as those from Carlson (1972), Stolarsky (1975), Mays (1983), or Bullen (2003). In Toader (1988), the integral mean value theorem for two functions is used, which seems to be more convenient than Cauchy's mean value theorem used in Bullen (2003). If p is a positive continuous function and $f \in CM(J)$, define a mean $V(f, p)$ by

$$V(f, p)(a, b) = f^{-1}\left(\frac{\int_a^b f(t)p(t)dt}{\int_a^b p(t)dt}\right). \tag{2.7}$$

The following special cases were already considered in Gini (1958):

$$V(e_n, p), V(\sin, p), V(\exp, p), V(e_1, pe_n),$$

where e_n is given by (2.5). Also, for $p(x) = 1$, $\forall x > 0$, the following means have been explicitly determined:

$$V(e_m, 1)(a, b) = \mathcal{D}_{m+1,1}(a, b) = \left[\frac{b^{m+1} - a^{m+1}}{(m+1)(b-a)}\right]^{\frac{1}{m}}, m \neq 0, -1,$$

known as the **Galvani mean**,

$$\mathcal{D}_{0,1}(a, b) = \mathcal{L}(a, b) = \frac{a - b}{\log a - \log b},$$

known as the **logarithmic mean**,

$$\mathcal{D}_{1,1}(a, b) = \mathcal{I}(a, b) = \frac{1}{e}\left(\frac{a^a}{b^b}\right)^{\frac{1}{a-b}},$$

referred to as the **identric mean**,

$$V(\sin, 1)(a, b) = \arcsin\left(\frac{\cos a - \cos b}{a - b}\right),$$

a trigonometric mean, or

$$V(\exp, 1)(a, b) = \log\left(\frac{e^a - e^b}{a - b}\right),$$

an exponential mean.

The special case $V(f, 1)$ in Berrone and Moro (1998) was named **Lagrangian mean**.

The well-known family of **extended means** (or **Stolarsky means** or **difference means**), defined in Stolarsky (1975), can be also obtained using the general formula (2.7). Taking

$$f = e_{r-s},\ p = p_{s-1},$$

where e_n is given by (2.5) and also $p_n = e_n$ for $n \neq 0$ but $p_0 = 1$, we obtain:
– the **extended mean** (for $r \cdot s \cdot (r - s) \neq 0$)

$$\mathcal{D}_{r,s}(a,b) = \left(\frac{s}{r} \cdot \frac{a^r - b^r}{a^s - b^s}\right)^{\frac{1}{r-s}};$$

– the **extended logarithmic mean** (for $r \neq 0,\ s = 0$)

$$\mathcal{L}_r(a,b) = \mathcal{D}_{r,0}(a,b) = \left(\frac{1}{r} \cdot \frac{a^r - b^r}{\log a - \log b}\right)^{\frac{1}{r}};$$

– the **extended identric** mean (for $r = s \neq 0$)

$$\mathcal{I}_r(a,b) = \mathcal{D}_{r,r}(a,b) = [\mathcal{I}(a^r, b^r)]^{\frac{1}{r}};$$

– the geometric mean (for $r = s = 0$)

$$\mathcal{G}(a,b) = \mathcal{D}_{0,0}(a,b) = \sqrt{ab}.$$

Of course, the power mean is the special case

$$\mathcal{P}_r(a,b) = \mathcal{D}_{2r,r}(a,b).$$

The mean $\mathcal{D}_{\frac{3}{2},\frac{1}{2}}$ can be written as

$$\mathcal{D}_{\frac{3}{2},\frac{1}{2}}(a,b) = \frac{a + \sqrt{ab} + b}{3}$$

which is known as **Heron mean**. It was generalized in Jia and Cao (2003) as

$$\mathcal{K}_p(a,b) = \left(\frac{a^p + (ab)^{p/2} + b^p}{3}\right)^{\frac{1}{p}},\ p > 0,$$

called **power-Heron mean**. Denote $\mathcal{K}_1 = \mathcal{K}$. Of course we have

$$\mathcal{K}_p = \mathcal{D}_{\frac{3p}{2},\frac{p}{2}}.$$

Remark 17. We have to underline that by bringing together in this book so many means we had to change the traditional notation for some of them. Our option was to preserve the notation for the most known and most used of them.

Taking also in (2.7)

$$f = e_1, \; p = \exp,$$

we obtain the **exponential mean**

$$\mathcal{E}(a,b) = \frac{ae^a - be^b}{e^a - e^b} - 1, \tag{2.8}$$

which was defined otherwise in Toader (1988a). In Sándor and Toader (1990), it was remarked that

$$\mathcal{E} = \mathcal{I}(\exp).$$

Other exponential means were defined and studied for example in Sándor and Toader (2006, 2006a).

The definition of the mean $V(f,p)$ was modified in Toader and Sándor (2006) using two functions as above, but only one integral. For $f, p \in CM(J)$, define $N(f,p)$ by

$$N(f,p)(a,b) = f^{-1}\left(\int_0^1 (f \circ p^{-1})[t \cdot p(a) + (1-t) \cdot p(b)]dt\right). \tag{2.9}$$

We obtain a symmetric mean $N(f,p)$ on J. Making the change of the variable

$$t = [p(b) - s] \,/\, [p(b) - p(a)],$$

the simpler representation

$$N(f,p)(a,b) = f^{-1}\left(\int_{p(a)}^{p(b)} \frac{(f \circ p^{-1})(s)\, ds}{p(b) - p(a)}\right)$$

is obtained. The special case $f = e_1$,

$$N_p(a,b) = \int_{p(a)}^{p(b)} \frac{p^{-1}(s)\, ds}{p(b) - p(a)}, \; \forall a, b \in J,$$

was defined in Sándor (1997) as

$$N_p(a,b) = \int_a^b \frac{xp'(x)\, dx}{p(b) - p(a)}, \; \forall a, b \in J.$$

Let us consider some examples.

The Stolarsky mean $\mathcal{D}_{r,s}$ is $N(e_{r-s}, e_s)$. We have

$$N_{\exp} = \mathcal{E},$$

which is the exponential mean (2.8). We can also give a new exponential mean

$$N_{1/\exp}(a,b)=\frac{a\cdot e^b-b\cdot e^a}{e^b-e^a}+1,\ a,b\geq 0.$$

Some trigonometric means such as

$$N_{\sin}(a,b)=\frac{b\cdot\sin b-a\cdot\sin a}{\sin b-\sin a}-\tan\frac{a+b}{2},\ a,b\in[0,\pi/2],$$

$$N_{\arcsin}(a,b)=\frac{\sqrt{1-b^2}-\sqrt{1-a^2}}{\arcsin a-\arcsin b},\ a,b\in[0,1],$$

$$N_{\tan}(a,b)=\frac{b\cdot\tan b-a\cdot\tan a+\ln(\cos b/\cos a)}{\tan b-\tan a},\ a,b\in[0,\pi/2]$$

and

$$N_{\arctan}(a,b)=\frac{\ln\sqrt{1+b^2}-\ln\sqrt{1+a^2}}{\arctan b-\arctan a},\ a,b\geq 0,$$

can also be obtained.

2.1.5 Comparison of means

We write

$$M\leq N\ \text{(on the interval } J)$$

to denote

$$M(a,b)\leq N(a,b),\ \forall a,b\in J.$$

If the inequality is strict for $a\neq b$, we write

$$M<N\ \text{(on the interval } J).$$

We say that M is **comparable to** N if

$$M\leq N\ \text{or}\ N\leq M.$$

Comparisons for Stolarski and Gini means were studied in Neuman and Páles (2003) or Czinder and Páles (2006).

The following theorem was proved in Toader and Toader (2002b) (see also Toader and Toader, 2005).

Theorem 16. *Among the Greek means we have only the following inequalities*

$$\mathcal{H}\leq\mathcal{G}\leq\mathcal{A}\leq\mathcal{F}_6\leq\mathcal{F}_5\leq\mathcal{C}$$

$$\mathcal{H} \leq \mathcal{F}_9 \leq \mathcal{F}_{10}, \; \mathcal{F}_8 \leq \mathcal{F}_7 \leq \mathcal{F}_5 \leq \mathcal{C}$$
$$\mathcal{F}_8 \leq \mathcal{A} \leq \mathcal{F}_6 \leq \mathcal{F}_5 \leq \mathcal{C}$$

and

$$\mathcal{G} \leq \mathcal{F}_{10}.$$

It is well known (see Bullen, 2003) that the first inequalities extend also to

$$\mathcal{H} < \mathcal{G} < \mathcal{L} < \mathcal{I} < \mathcal{A}, \text{ on } (0, \infty). \tag{2.10}$$

The following lemma is proved in Sándor and Toader (1999).

Lemma 3. *If the function $f : \mathbb{R}_+ \to \mathbb{R}_+$ is strictly monotonous, the function $g : \mathbb{R}_+ \to \mathbb{R}_+$ is strictly increasing, and the composed function $g \circ f^{-1}$ is convex, then the inequality*

$$V(f, p) < V(g, p)$$

holds for every positive function p, where $V(f, p)$ is given by (2.7).

The means $\mathcal{A}$, $\mathcal{G}$ and $\mathcal{L}$ can be obtained as the means $V(e_n, 1)$ for $n = 1, n = -2$, respectively $n = -1$. So the relations between them follow from the above result. But $\mathcal{H} = V(e_1, e_{-3})$, thus the inequality $\mathcal{H} < \mathcal{G}$ cannot be proved this way.

The result from the previous lemma holds also in the case of the mean $N(f, p)$ defined by (2.9). As it was proved in Toader and Sándor (2006), using a simplified variant of the integral inequality of Jensen for the convex function $g \circ f^{-1}$ (see Bullen, 2003), we have

$$(g \circ f^{-1}) \left(\int_0^1 (f \circ p^{-1}) \left[t \cdot p(a) + (1-t) \cdot p(b) \right] dt \right)$$
$$\leq \int_0^1 (g \circ f^{-1}) \circ (f \circ p^{-1}) \left[t \cdot p(a) + (1-t) \cdot p(b) \right] dt,$$

and applying the increasing function g^{-1} we get the desired inequality.

A similar result of monotony with respect to the function p was also proved in this case.

Theorem 17. *If p is a strictly monotone real function on J and q is a strictly increasing real function on J, such that $q \circ p^{-1}$ is strictly convex, then*

$$N(f, p) < N(f, q) \text{ on } J,$$

for each strictly monotonous function f.

Corollary 5. *If the function q is strictly convex and strictly increasing then*

$$N_q > \mathcal{A}.$$

Remark 18. If we replace the convexity by the concavity and/or the increase by the decrease in the above theorems, the same/the opposite inequalities are obtained.

Thus we have

$$\mathcal{L}, N_{\sin}, N_{\arctan} < \mathcal{A},$$

but

$$\mathcal{E}, N_{\arcsin}, N_{\tan} > \mathcal{A}.$$

More generally, from the second theorem we deduce that for $m \cdot n > 0$ we have

$$N_{e_n} < N_{e_m}, \text{ if } n < m.$$

So

$$\begin{aligned} N_{e_n} &> \mathcal{A}, \text{ for } n > 1, \\ \mathcal{L} < N_{e_n} &< \mathcal{A}, \text{ for } 0 < n < 1, \\ \mathcal{G} < N_{e_n} &< \mathcal{L}, \text{ for } -1/2 < n < 0, \\ \mathcal{H} < N_{e_n} &< \mathcal{G}, \text{ for } -2 < n < -1/2, \end{aligned}$$

and

$$N_{e_n} < \mathcal{H}, \text{ for } n < -2.$$

To present the next results, we need some definitions. Let F_p, $p \in \mathbb{R}$, be a family of means. It is said to be an **increasing family** if

$$F_p < F_q \text{ for } p < q.$$

A **lower estimation** of a given mean M by this family of means assumes the determination of some index p such that $F_p < M$. The **optimal lower estimation** is given by the mean F_r where r is the greatest index p such that $F_p < M$. Similarly the notions of **upper estimation** and **optimal upper estimation** are defined.

The most known and used increasing family of means is that of power means. The first determination of lower and upper optimal estimations by power means was done for the logarithmic mean in Lin (1974):

$$\mathcal{G} = \mathcal{P}_0 < \mathcal{L} < \mathcal{P}_{1/3}. \tag{2.11}$$

It was generalized in Pittenger (1980) as follows: for each $r \neq 0$, we have

$$\mathcal{P}_{\min\{r_1,r_2\}} < \mathcal{D}_{r,1} < \mathcal{P}_{\max\{r_1,r_2\}},$$

where

$$r_1 = \frac{r+1}{3}$$

and

$$r_2 = \begin{cases} \frac{r-1}{\log_2 r} & \text{for } r > 0, r \neq 1 \\ \ln 2 & \text{for} \quad r = 1 \\ 0 & \text{for} \quad r < 0 \end{cases}.$$

For example, for $r = 1$ we have

$$\mathcal{P}_{2/3} < \mathcal{I} < \mathcal{P}_{\ln 2}.$$

For $r = -1, 1/2$ or 2 we have equality, thus

$$\mathcal{D}_{-1,1} = \mathcal{G} = \mathcal{P}_0, \; \mathcal{D}_{1/2,1} = \mathcal{P}_{1/2}, \; \mathcal{D}_{2,1} = \mathcal{A} = \mathcal{P}_1.$$

After all (2.10) it can be refined as

$$\mathcal{H} < \mathcal{G} < \mathcal{L} < \mathcal{P}_{1/3} < \mathcal{P}_{2/3} < \mathcal{I} < \mathcal{P}_{ln2} < \mathcal{A} < \mathcal{E}\,, \text{ on } (0, \infty). \qquad (2.12)$$

The estimation by power means can be unsatisfactory. That is why other families of means were also used. In fact, for the estimation of the logarithmic mean, even before using the power means, in Karamata (1960) there were used the means defined by

$$\frac{ab^p + ba^p}{a^p + b^p} = \mathcal{S}_{1-p,-p}(a,b) = \mathcal{C}_{1-p}(a,b).$$

It was proved that

$$\mathcal{C}_{2,3} < \mathcal{L},$$

and in Alzer (1993) it was shown that this estimation is optimal. An upper optimal estimation by power-type Heron means was given in Jia and Cao (2003):

$$\mathcal{L} < \mathcal{K}_{1/2}.$$

The optimal evaluation by the family of special Gini means S_p was established in Costin and Toader (2013b), where the following result is being proved.

Theorem 18. *The optimal evaluation of the logarithmic mean $\mathcal{L}$ by means from the families $\mathcal{P}_p$, $\mathcal{K}_p$, $\mathcal{S}_p$ and $\mathcal{C}_p$ can be ordered as follows:*

$$\mathcal{G} = \mathcal{P}_0 = \mathcal{K}_0 < \mathcal{S}_{1/3} < \mathcal{C}_{2/3} < \mathcal{L} < \mathcal{K}_{1/2} < \mathcal{P}_{1/3} < \mathcal{S}_1 = \mathcal{C}_1 = \mathcal{A}. \quad (2.13)$$

2.1.6 Weighted means

The **weighted** generalization of some means is also well known. For instance, the quasi-arithmetic means can be generalized in the following way: let $f : J \to \mathbb{R}$ be a strictly monotonic function and $g, h : J^2 \to \mathbb{R}_+$ be two positive functions; define a mean by

$$W_{f,g,h}(a,b) = f^{-1}\left(\frac{f(a)g(a,b) + f(b)h(a,b)}{g(a,b) + h(a,b)}\right); a, b \in J.$$

For example, we can consider means like

$$T_n(a,b) = \left[a^n \cos^2(a+b) + b^n \sin^2(a+b)\right]^{1/n}, n \neq 0$$

and

$$T_0(a,b) = \lim_{n \to 0} T_n(a,b) = a^{\cos^2(a+b)} b^{\sin^2(a+b)}.$$

The special case

$$W_{f,g}(a,b) = f^{-1}\left(\frac{f(a)g(a) + f(b)g(b)}{g(a) + g(b)}\right); a, b \in J,$$

was defined in Bajraktarević (1958) and is called **Bajraktarević mean**, while $A_\lambda(f)$, defined by

$$\mathcal{A}_\lambda(f)(a,b) = f^{-1}(\lambda f(a) + (1-\lambda) f(b)); a, b \in J,$$

is called **weighted quasi-arithmetic mean**.

For example, the weighted power means $\mathcal{P}_{n,\lambda} = A_\lambda(e_n)$ are given by

$$\mathcal{P}_{n;\lambda}(a,b) = \begin{cases} [\lambda \cdot a^n + (1-\lambda) \cdot b^n]^{1/n}, & n \neq 0 \\ a^\lambda \cdot b^{1-\lambda} & , n = 0 \end{cases},$$

with $\lambda \in [0,1]$ fixed. Of course, for $\lambda = 0$ or $\lambda = 1$, we have

$$\mathcal{P}_{n;0} = \Pi_2 \text{ respectively } P_{n;1} = \Pi_1, \ \forall n \in \mathbb{R},$$

where the projections Π_1 and Π_2 are defined by (2.4). The weighted arithmetic, geometric, and harmonic means, denoted by A_λ, $\mathcal{G}_\lambda$, and $\mathcal{H}_\lambda$, are obtained for $n = 1, 0$, or -1.

Weighted Gini means are defined by

$$\mathcal{S}_{r,s;\lambda}(a,b)=\left[\frac{\lambda\cdot a^r+(1-\lambda)\cdot b^r}{\lambda\cdot a^s+(1-\lambda)\cdot b^s}\right]^{\frac{1}{r-s}},\ r\neq s$$

and

$$\mathcal{S}_{r,r;\lambda}(a,b)=\left[a^{\lambda\cdot a^r}b^{(1-\lambda)\cdot b^r}\right]^{\frac{1}{\lambda\cdot a^r+(1-\lambda)\cdot b^r}}$$

with $\lambda\in[0,1]$ fixed. **Weighted Lehmer means**, $\mathcal{C}_{r;\lambda}=\mathcal{S}_{r,r-1;\lambda}$, are also used. We can remark that again,

$$\mathcal{S}_{r,s;0}=\mathcal{C}_{r;0}=\Pi_2 \text{ and } \mathcal{S}_{r,s;1}=\mathcal{C}_{r;1}=\Pi_1.$$

We obtained previously weighted variants for the first four Greek means. For the last six Greek means, weighted variants were defined in Toader (2005), using a modified method of proportions. Consider a set of three numbers, $a>m>b>0$. Remember that the arithmetic mean is defined by the proportion

$$\frac{a-m}{m-b}=\frac{a}{a}.$$

Take $\lambda\in(0,1)$ and multiply the first member of the proportion by $\lambda/(1-\lambda)$. Now m will give the weighted arithmetic mean of a and b. We shall proceed like this in all cases. Remark that in the case of the geometric mean, a weighted variant is obtained:

$$\mathcal{G}(\lambda,a,b)=\frac{1}{2\lambda}\left[\sqrt{(1-2\lambda)^2\cdot a^2+4\lambda(1-\lambda)ab}-(1-2\lambda)a\right],$$

which is completely different from the usual

$$\mathcal{G}_\lambda(a,b)=a^\lambda\cdot b^{1-\lambda}.$$

The following method of construction of new means as an integral average of a given family of means was considered in Toader (1998a).

Lemma 4. *If* $\mathbf{R}=\{R_t,\ t\in[\alpha,\beta]\}$ *is a family of means and* $p:\mathbb{R}_+\to\mathbb{R}$ *is a strictly monotonic function, then*

$$M_{p,\mathbf{R}}(a,b)=p^{-1}\left(\frac{1}{\beta-\alpha}\int_\alpha^\beta p(R_t(a,b))dt\right)\tag{2.14}$$

defines a mean.

The representation of $\mathcal{A} \otimes \mathcal{G}$ by (1.16) can be interpreted this way. Let us consider another special case of the above construction of means. Denote by $\mathbf{P}_n = \{\mathcal{P}_{n,t}\,,\ t \in [0,1]\}$ a family of weighted power means. The following result holds.

Lemma 5. *We have*

$$M_{e_r,\mathbf{P}_n} = \mathcal{D}_{r+n,n} \tag{2.15}$$

for every $r, n \in \mathbb{R}$.

Proof. It is enough to make the computations indicated by the definition (2.14) in the following cases: $i)$ $n \neq 0,\ n \neq -r$; $ii)$ $n = 0,\ r \neq 0$; $iii)$ $n = -r \neq 0$; $iv)$ $n = r = 0$. □

Remark 19. If we write (2.15) as

$$\mathcal{D}_{r,s} = M_{e_{r-s},\mathbf{P}_s},$$

we get an **integral representation** of the Stolarsky means. This result appears in Carlson (1972) for $r = -1, s = 1$, in Neuman (1994) for $r = 1, s = 0$, and in Pittenger (1980) for $s = 1$ with arbitrary r.

The means $M_{p,\mathbf{R}}$ defined by (2.14), for different functions p or different families of means $\mathbf{R}$, are also compared in Toader (1998a).

Theorem 19. *If the function $q \circ p^{-1}$ is convex and q^{-1} is increasing, then*

$$M_{p,\mathbf{R}} < M_{q,\mathbf{R}} \tag{2.16}$$

for every family of means $\mathbf{R}$.

Proof. Jensen's inequality for $q \circ p^{-1}$ gives

$$q \circ p^{-1}\left(\frac{1}{\beta-\alpha}\int_\alpha^\beta p(R_t(a,b))dt\right) \leq \frac{1}{\beta-\alpha}\int q \circ p^{-1} \circ p(R_t(a,b))dt,$$

and applying q^{-1} we get the result. □

Remark 20. For other combinations of convexity/concavity of $q \circ p^{-1}$ and monotonicity of q^{-1} we also get (2.16) or its reverse.

Theorem 20. *If* $\mathbf{R}$ *and* $\mathbf{R}'$ *are two families of means such that $R_t < R'_t$, for all $t \in [\alpha, \beta]$, then*

$$M_{p,\mathbf{R}} < M_{p,\mathbf{R}'} \tag{2.17}$$

for every function p.

Corollary 6. *The means $\mathcal{D}_{r,s}$ are increasing in both r and s.*

Remark 21. This property is known from Stolarsky (1975). A more general problem related to the comparability of two arbitrary means $\mathcal{D}_{r,s}$ and $\mathcal{D}_{r',s'}$ is solved in Leach and Sholander (1983) and Páles (1988a) for $r \neq s$ and $r' \neq s'$.

In Czinder and Páles (2003), the final result allowing the case of equal parameters is proved.

Theorem 21. *The comparison inequality*

$$\mathcal{D}_{r,s} \leq \mathcal{D}_{r',s'}$$

holds if and only if the conditions

$$r + s \leq r' + s'$$

and

$$l(r,s) \leq l(r',s'),\ k(r,s) \leq k(r',s')$$

are satisfied, where

$$l(r,s) = \begin{cases} \mathcal{L}(r,s), & \text{if } r,s > 0 \\ 0, & \text{otherwise} \end{cases}$$

and

$$k(r,s) = \begin{cases} \dfrac{|r| - |s|}{r - s}, & \text{if } \ r \neq s \\ sign(r), & \text{if } \ r = s. \end{cases}$$

Remark 22. Similarly, two sum means are compared in Páles (1988) for distinct parameters, and in Czinder and Páles (2000) in the general case.

Theorem 22. *The comparison inequality*

$$\mathcal{S}_{r,s} \leq \mathcal{S}_{r',s'}$$

holds if and only if

$$r + s \leq r' + s'$$

and

$$\begin{cases} \min\{r,s\} \leq \min\{r',s'\}, & \text{if } \ \min\{r,s,r',s'\} \geq 0 \\ \max\{r,s\} \leq \max\{r',s'\}, & \text{if } \ \max\{r,s,r',s'\} \leq 0 \\ k(r,s) \leq k(r',s'), & \text{if } \ \min\{r,s,r',s'\} < 0 < \max\{r,s,r',s'\}. \end{cases}$$

2.1.7 Weak and angular inequalities

A comparability of means on a subset was considered in Schoenberg (1982) and Foster and Phillips (1985).

Definition 5. The means M and N defined on J are in the relation

$$M \leq_D N,$$

where $D \subset J \times J$, if

$$M(a,b) \leq N(a,b),\ \forall (a,b) \in D.$$

If the last inequality is strict for $a \neq b$, we write

$$M <_D N.$$

If $M <_D N$ and $N \leq_{D'} M$, where $D' = J \times J \setminus D$, we write

$$M \prec_D N.$$

Remark 23. If the means M and N are symmetric and $M \prec_D N$, then D has a kind of **symmetry**:

$$(a,b) \in D \Rightarrow (b,a) \in D.$$

The following definition was given in Toader (1987).

Definition 6. The means M and N defined on J are in the **weak relation**

$$M \prec N$$

if $M \prec_D N$ for $D = \{(x,y) \in J \times J; x < y\}$.

Remark 24. If $p < q$, we have

$$\mathcal{A}_q \prec \mathcal{A}_p,\ \mathcal{G}_q \prec \mathcal{G}_p,\ \mathcal{H}_q \prec \mathcal{H}_p.$$

Remark 25. We have seen that homogeneous means are defined on $(0, \infty)$. The comparison of homogeneous means can be done only on special sets.

Definition 7. The set $D \subset \mathbb{R} \times \mathbb{R}$ is called **starshaped** if

$$(a,b) \in D, t > 0 \text{ implies } (ta, tb) \in D.$$

The following property holds.

Lemma 6. *If the means M and N are homogeneous and $M \prec_D N$, then the set D is starshaped.*

A relation of this kind was given in Toader and Toader (2002b). Let $m > 1$.

Definition 8. The means M and N are in the **symmetric angular relation**

$$M \prec_m N$$

if $M \prec_D N$ for $D = \{(x, y) \in \mathbb{R}_+^2;\ y/m < x < my\}$.

Theorem 23. *Let us make the following denotations:*

$$s1 = \frac{1+\sqrt{5}}{2},\ s2 = \frac{2+\sqrt{2}}{2},\ s3 = \frac{3+\sqrt{5}}{2},$$

and $t1, t2, t3, t4, t5 > 1$ *be the corresponding roots of the equations*

$$t^3 - 2t^2 + t - 1 = 0,\ t^3 - t^2 - 2t + 1 = 0,\ t^3 - t^2 - t - 1 = 0,$$
$$t^3 - 3t^2 + 2t - 1 = 0,\ \textit{respectively}\ t^3 - 5t^2 + 4t - 1 = 0.$$

We have the following symmetric angular relations between the Greek means:

$$\mathcal{F}_5 \prec_2 \mathcal{F}_{10},\ \mathcal{F}_8 \prec_2 \mathcal{H},\ \mathcal{F}_7 \prec_2 \mathcal{A},\ \mathcal{A} \prec_2 \mathcal{F}_9,\ \mathcal{A} \prec_3 \mathcal{F}_{10},$$
$$\mathcal{F}_7 \prec_{s1} \mathcal{H},\ \mathcal{C} \prec_{s1} \mathcal{F}_9,\ \mathcal{F}_5 \prec_{s2} \mathcal{F}_9,$$
$$\mathcal{F}_8 \prec_{s3} \mathcal{G},\ \mathcal{G} \prec_{s3} \mathcal{F}_9,\ \mathcal{F}_7 \prec_{t1} \mathcal{G},\ \mathcal{F}_6 \prec_{t2} \mathcal{F}_9,$$
$$\mathcal{C} \prec_{t3} \mathcal{F}_{10},\ \mathcal{F}_7 \prec_{t4} \mathcal{F}_6,\ \mathcal{F}_6 \prec_{t4} \mathcal{F}_{10}\ \textit{and}\ \mathcal{F}_8 \prec_{t5} \mathcal{F}_{10}.$$

Some inequalities between trigonometric means are proved similarly, in Jecklin (1953). In Toader and Toader (2008) and Toader (1989, 1998c) are proved some weak and angular inequalities between weighted geometric means or some non-symmetric means.

2.1.8 Operations with means

Ordinary denotations for operations with functions are used in what follows. For example, $M + N$ is defined by

$$(M + N)(a, b) = M(a, b) + N(a, b),\ \forall a, b \in J,$$

while λM by

$$(\lambda M)(a, b) = \lambda \cdot M(a, b),\ \forall a, b \in J.$$

Of course, if M and N are means, the result of the operation with functions is not a mean. We have to combine more operations with functions to get a (partial) operation with means. For instance, in Tricomi (1970) the linear combinations

$$r\mathcal{A} + s\mathcal{G} + (1 - r - s)\mathcal{H}$$

are characterized, which are means. In Janous (2001) a **generalized Heron mean** $\lambda\mathcal{G} + (1 - \lambda)\mathcal{A}$ for $\lambda \in (0, 1)$ was considered. For $\lambda = 1/3$ we get the Heron mean.

Of course, the set of means (defined on an interval J) is convex, i.e. $\lambda M + (1 - \lambda)N$ is a mean for $\lambda \in [0, 1]$. The following properties are given in Borwein and Borwein (1987).

Proposition 2. *The isotone (symmetric), (strict), (homogeneous) means form a convex set.*

Remark 26. Using operations with $\wedge$ and $\vee$, we can give the Greek means as follows:

$$\mathcal{A} = \frac{\vee + \wedge}{2},\ \mathcal{G} = \sqrt{\vee\wedge},\ \mathcal{H} = \frac{2\vee\wedge}{\vee + \wedge},\ \mathcal{C} = \frac{\vee^2 + \wedge^2}{\vee + \wedge},$$

$$\mathcal{F}_5 = \frac{1}{2}\left[\vee - \wedge + \sqrt{(\vee - \wedge)^2 + 4\wedge^2}\right],$$

$$\mathcal{F}_6 = \frac{1}{2}\left[\wedge - \vee + \sqrt{(\vee - \wedge)^2 + 4\vee^2}\right],$$

$$\mathcal{F}_7 = \frac{\vee^2 - \vee\wedge + \wedge^2}{\vee},\ \mathcal{F}_8 = \frac{\vee^2}{2\vee - \wedge},$$

$$\mathcal{F}_9 = \frac{\wedge(2\vee - \wedge)}{\vee} \text{ and } \mathcal{F}_{10} = \frac{1}{2}\left[\wedge + \sqrt{\wedge(4\vee - 3\wedge)}\right].$$

Of great importance for what follows is the **composition** of means. Given three means M, N, and P on the same interval J, the expression

$$P(M, N)(a, b) = P(M(a, b), N(a, b)),\ \forall a, b \in J$$

also defines a mean $P(M, N)$ on J. The special cases $\wedge(M, N)$ and $\vee(M, N)$ are denoted simpler as $M \wedge N$ and $M \vee N$ respectively.

Some relations like

$$\mathcal{G}(\mathcal{P}_n, \mathcal{P}_{-n}) = \mathcal{G},$$

or

$$\mathcal{P}_{2n}(\mathcal{G}, \mathcal{P}_{2n}) = \mathcal{P}_n$$

are given in Gini (1958).

The following pre-means were considered in Kim (1999):

$$K_1 = \mathcal{A}\left(\frac{\mathcal{G}}{\mathcal{P}_2}\right)^2, \; K_2 = \mathcal{A}\left(\frac{\mathcal{G}}{\mathcal{O}}\right)^2, \; K_3 = \mathcal{A}\frac{\mathcal{P}_2}{\mathcal{G}}, \; K_4 = \mathcal{A}\left(\frac{\mathcal{P}_2}{\mathcal{G}}\right)^2, \tag{2.18}$$

$$K_5 = \mathcal{A}\left(\frac{\mathcal{A}}{\mathcal{G}}\right)^2, \; K_6 = \mathcal{A}\frac{\mathcal{G}}{\mathcal{P}_2}, \text{ and } K_7 = \mathcal{A}\left(\frac{\mathcal{O}}{\mathcal{G}}\right)^2, \tag{2.19}$$

where $\mathcal{O}$ is given by

$$\mathcal{O} = \mathcal{P}_{4;3/4}(\mathcal{P}_4, \mathcal{G}).$$

In Toader (2001), it was proved that only K_1 and K_6 are means. In fact, the following problem was studied: if M, N, and P are means, let us consider the pre-mean

$$\frac{M \cdot N}{P}.$$

For what triples of means we get a mean? The following results were proved:

Lemma 7. *If the means M, N, and P are in the relation $M \leq N \leq P$, then MP/N is a mean.*

Lemma 8. *If the means M and N are such that $M, N \geq \mathcal{G}$ then $M\mathcal{G}/N$ is a mean.*

Other results of this type were given in Toader and Toader (2002):

Theorem 24. *For $m, n, q \geq 0$, the pre-mean $\frac{\mathcal{P}_n \mathcal{P}_m}{\mathcal{P}_q}$ is a mean if and only if the condition*

$$q \geq \frac{nm}{n+m}$$

is satisfied.

Corollary 7. *The pre-mean $\mathcal{I}^2/\mathcal{P}_n$ is a mean if and only if $n \geq \ln\sqrt{2}$ while the pre-mean $\mathcal{P}_n^2/\mathcal{I}$ is a mean if and only if $n \leq \ln 4$.*

Corollary 8. *The pre-mean $\mathcal{P}_n^2/\mathcal{E}$ is a mean if and only if $n \leq 2$ but there is no $n > 0$ such that $\mathcal{E}^2/\mathcal{P}_n$ be a mean.*

Corollary 9. *The pre-mean $\mathcal{L}^2/\mathcal{P}_n$ is a mean if $n \geq 1/6$ but there is no $n > 0$ such that $\mathcal{P}_n^2/\mathcal{L}$ be a mean.*

Relations among the parameters $m, n, q, r \geq 0$ such that the expression

$$\mathcal{P}_m \cdot (\mathcal{P}_n/\mathcal{P}_q)^r$$

is a mean were found in Toader and Toader (2003).

2.1.9 Universal means

The following definition was introduced in Toader (2007a).

Definition 9. A mean U is called **upper universal** if there exists a constant $p > 0$ such that

$$p\vee \leq U \leq \vee.$$

Remark 27. Of course U is an upper universal mean if and only if the inequality

$$M \leq \frac{1}{p} U$$

holds for every mean M.

Theorem 25. *The following means are upper universal: i)* $\mathcal{S}_{r,s;\lambda}$ *for* $r > s > 0$*; ii)* $\mathcal{S}_{r,r;\lambda}$ *for* $r > 0$*; iii)* $\mathcal{C}_{r;\lambda}$ *for* $r > 1$*; iv)* $\mathcal{P}_{r;\lambda}$ *for* $r > 0$*; v)* $\mathcal{D}_{r,s}$ *for* $rs(r-s) \neq 0$*; vi)* $\mathcal{F}_5$*; vii)* $\mathcal{F}_6$*; viii)* $\mathcal{F}_7$*; ix)* $\mathcal{F}_8$*.*

General results of the following types were also proved.

Theorem 26. *If* $f : \mathbb{R}_+ \to \mathbb{R}$ *is a bijective function and* f^{-1} *is concave, then the weighted quasi arithmetic mean* $\mathcal{A}_\lambda(f)$ *is upper universal.*

Theorem 27. *If* $f : \mathbb{R}_+ \to \mathbb{R}$ *is an increasing convex function, then*

$$V(f,1)(a,b) = f^{-1}\left(\frac{1}{b-a}\int_a^b f(x)dx\right), \forall a,b > 0,$$

defines an upper universal mean $V(f,1)$*.*

Definition 10. A mean U is called **lower universal** if there exists a constant p such that

$$\wedge \leq U \leq p\wedge.$$

Remark 28. The mean U is lower universal if and only if for some $p \neq 0$ the inequality

$$M \geq \frac{1}{p} U$$

holds for every mean M.

Theorem 28. *The following means are lower universal: i)* $\mathcal{S}_{r,s;\lambda}$ *for* $r < s < 0$*; ii)* $\mathcal{S}_{r,r;\lambda}$ *for* $r < 0$*; iii)* $\mathcal{P}_{r;\lambda}$ *for* $r < 0$*; iv)* $\mathcal{F}_9$*.*

Definition 11. A mean U is called **universal** if it is upper universal or lower universal.

Remark 29. If U is an universal mean then there exists a constant p such that each mean is comparable with pU.

Theorem 29. *The following means are not universal: i) the Gini mean* $\mathcal{S}_{r,s;\lambda}$ *for* $s < 0 < r$*; ii) the logarithmic mean* $\mathcal{L}_r$, $r \neq 0$*; iii) the geometric mean* $\mathcal{G}$*; iv) the Greek mean* $\mathcal{F}_{10}$.

Remark 30. If U is an upper (lower) universal mean and $W \geq qU$ (respectively $W \leq qU$), then W is also an upper (respectively lower) universal mean.

Example 2. The inequalities (2.12) imply that the means $\mathcal{I}$ and $\mathcal{E}$ are upper universal.

Remark 31. In the sequence of inequalities (2.12) the first mean $\mathcal{H}$ is lower universal, then $\mathcal{G}$ and $\mathcal{L}$ are not universal, while the last means are upper universal. In fact this order cannot be changed in no chain of inequalities.

2.1.10 Invariant means

As it was shown in Chapter 1, the determination of the $\mathcal{AGM}$ was done by Gauss with the help of the relation (1.18), thus

$$I(\mathcal{A}(a,b), \mathcal{G}(a,b)) = I(a,b), \forall a,b > 0,$$

leading to

$$\mathcal{M}\left(\mathcal{A}, \mathcal{G}\right) = \mathcal{M}.$$

This property was generalized in Borwein and Borwein (1987) as to the following.

Definition 12. The mean P is called (M,N)-**invariant** (or invariant with respect to M and N) if it verifies

$$P(M,N) = P.$$

Obviously we have the following **duality property**:

Lemma 9. *If the symmetric mean* P *is* (M,N)*-invariant, then it is also* (N,M)*-invariant.*

The following property was proved in Toader and Toader (2006).

Lemma 10. *If the means M and N are symmetric and P is (M,N)-invariant, then P is also symmetric.*

A similar result can be also proved.

Lemma 11. *If the means P and M are symmetric, P is strictly isotonic and (M,N)-invariant, then N is also symmetric.*

Proof. We have

$$P(M(a,b),N(a,b)) = P(a,b),\ P(M(b,a),N(b,a)) = P(b,a), \forall a,b \in J.$$

As P and M are symmetric, the second equality gives

$$P(M(a,b),N(b,a)) = P(a,b), \forall a,b \in J,$$

thus

$$P(M(a,b),N(a,b)) = P(M(a,b),N(b,a)), \forall a,b \in J.$$

The strict isotony of P implies the symmetry of N. □

The importance of this notion in the study of double sequences will be discussed in the next chapter. For the moment, let us underline some problems, interesting in themselves. Given a family $\mathcal{Z}$ of means, a first problem is that of the study of **invariance** of a given mean P **with respect to the family** $\mathcal{Z}$. This means the determination of all the pairs of means (M,N) from $\mathcal{Z}$ such that P is (M,N)-invariant. A second problem is named **invariance in the family** $\mathcal{Z}$. It consists in determining all the triples of means (P,M,N) from $\mathcal{Z}$ such that P is (M,N)-invariant. A third type of problems was called in Brenner and Mays (1987) as **reproducing identities** and assumes the determination of quadruples of means (P,M,N,Q) from $\mathcal{Z}$ such that

$$P(M,N) = Q.$$

This problem has the trivial solution

$$P(M,M) = M$$

and all the solutions of the invariance problem

$$P(M,N) = P,$$

but it can also have other solutions.

Many such problems were formulated as functional equations. The first one was related to the invariance of arithmetic mean $\mathcal{A}$ with respect to the family of quasi-arithmetic means $\mathcal{A}(f)$. It was solved in Sutô (1914, 1914a) for analytic functions f and in Matkowski (1999) for the second order continuously differentiable functions f. As a consequence, this problem is called the **Matkowski–Sutô problem**. The regularity assumptions were weakened step-by-step in Daróczy et al. (2005), Daróczy and Páles (2001, 2002), arriving at simple continuity hypothesis on the functions f. The problem of invariance of the arithmetic mean $\mathcal{A}$ was studied later:

– with respect to the family of Lagrangian means, in Matkowski (2005);

– with respect to the family of Beckenbach–Gini means, in Domsta and Matkowski (2006);

– with respect to the family of weighted quasi-arithmetic means $\mathcal{A}_\lambda(f)$, in Burai (2007);

– with respect to the family of generalized quasi-arithmetic means $\mathcal{A}^{[f,g]}$, in Baják and Páles (2009), where

$$\mathcal{A}^{[f,g]}(a,b) = (f+g)^{-1}\left(f(a)+g(b)\right);$$

– with respect to the family of Lagrangian quasi-arithmetic means $\mathcal{A}_{[\mu]}^{[f]}$, in Makó and Páles (2009), where

$$\mathcal{A}_{[\mu]}^{[f]}(a,b) = f^{-1}\left(\int_0^1 f\left(tx+(1-t)\,y\right)d\mu(t)\right);$$

– with respect to the family of Bajraktarević means $W_{f,g}$, in Jarczyk (2010).

Given a mean M on J, a continuous strictly monotonic function f on J, and the numbers $p,q \in [0,1]$, in Daróczy and Páles (2001) a M**-conjugate** mean $M_{(p,q)}^{f}$ is defined by the formula

$$M_{(p,q)}^{f}(a,b) = f^{-1}(pf(a)+qf(b)+(1-p-q)f(M(a,b))),\ a,b \in J.$$

The family of means $\{M_{(p,q)}^{f} : p,q \in [0,1]\}$ contains the mean $M = M_{(0,0)}^{f}$, the weighted quasi-arithmetic mean $A_p(f) = M_{(p,1-p)}^{f}$, and the mean $M_{(1,1)}^{f}$ given by

$$M_{(1,1)}^{f}(a,b) = f^{-1}(f(a)+f(b)-f(M(a,b))),\ a,b \in J.$$

This can be written as

$$f(M(a,b)) + f(M_{(1,1)}^{f}(a,b)) = f(a)+f(b),$$

or

$$\mathcal{A}(f)(M, M^f_{(1,1)}) = \mathcal{A}(f),$$

thus $\mathcal{A}(f)$ is $(M, M_{(1,1)})$-invariant. An extension theorem for a Matkowski–Sutô problem was given in Daróczy et al. (2003).

Also in Głazowska and Matkowski (2007) the problem of invariance of the geometric mean $\mathcal{G}$ with respect to the family of Lagrangian means was studied.

The problem of invariance was studied in the family of Beckenbach–Gini means in Matkowski (2002) and in the family of weighted quasi-arithmetic means in Jarczyk and Matkowski (2006) and Jarczyk (2009). The second problem was solved in Jarczyk (2007) as follows.

Theorem 30. *The functions $\alpha, \beta, \gamma \in CM(J)$ and the numbers $p, q, r \in (0, 1)$ satisfy*

$$\mathcal{A}_p(\alpha)\left(\mathcal{A}_q(\beta), \mathcal{A}_r(\gamma)\right) = \mathcal{A}_p(\alpha)$$

if and only if the following two conditions are fulfilled: (i) $p = r/(1 - q + r)$; (ii) there exist $a, c \in \mathbb{R}\backslash\{0\}$ and $b, d \in \mathbb{R}$ such that

$$\beta(x) = a \cdot \alpha(x) + b, \gamma(x) = c \cdot \alpha(x) + d, x \in J,$$

or $p = 1/2$ and

$$\beta(x) = a \cdot e^{\lambda\alpha(x)} + b, \gamma(x) = c \cdot e^{-\lambda\alpha(x)} + d, x \in J,$$

with some $\lambda \in \mathbb{R}\backslash\{0\}$.

Remark 32. For $p = 1/2$ and $\alpha = e_1$, we obtain the solution of the Matkowski–Sutô problem.

Remark 33. Let $J \subset \mathbb{R}$ be an interval, $p, r \in (0, 1)$ be fixed, and the functions $f, g : J \to (0, \infty)$ be continuous, f being strictly increasing and g being strictly decreasing, and $\frac{f}{g}$ be one-to-one. Then the following conditions are equivalent:

(i) the quotient mean $Q^{[f,g]}$ is $(\mathcal{A}^p_f, \mathcal{A}^r_g)$-invariant;

(ii) the product fg is a constant function and $p + r = 1$;

(iii) for all $a, b \in J$,

$$Q^{[f,g]}(a, b) = f^{-1}\left(\sqrt{f(a)f(b)}\right),$$

$$\mathcal{A}_p(f)(a, b) = f^{-1}\left(\frac{f(a)f(b)}{pf(a) + (1 - p)f(b)}\right).$$

The problem of invariance in the class of Heinz means was solved in Besenyei (2012). It was proved that if $0 < p, q, r \leq \frac{1}{2}$ then the invariance equation

$$\mathcal{V}_{p,1-p}(\mathcal{V}_{q,1-q}, \mathcal{V}_{r,1-r}) = \mathcal{V}_{p,1-p}$$

is valid if and only if $p = q = r$.

The first reproducing identities problem was studied in Brenner and Mays (1987) for the families of Lehmer means and for that of power means. Their results will be given later. A problem which is somehow related was solved in Daróczy and Páles (2003). It is proved that the solutions of the equation

$$\mathcal{A}_{\mu}\left(\mathcal{A}_{\lambda}(\alpha), \mathcal{A}_{\lambda}(\beta)\right) = \mathcal{A}_{\lambda}$$

on an interval J are the following:

(i) If $\lambda \neq \frac{1}{2}$, then $\alpha = \beta = e_1$;

(ii) If $\lambda = \frac{1}{2}$ and $\mu \notin \{\frac{1}{2}, 2\}$, then $\alpha = \beta = e_1$;

(iii) If $\lambda = \mu = \frac{1}{2}$, then $\alpha(x) = \exp(px)$ and $\beta(x) = \exp(-px)$, for some $p > 0$;

(iv) If $\lambda = \frac{1}{2}$ and $\mu = 2$, then $\alpha = \beta = e_1$; or

$\alpha(x) = \sqrt{p+x}$, $\beta(x) = \log\sqrt{p+x}$, with such a p that $p + x > 0$ for $x \in J$; or

$\alpha(x) = \sqrt{p-x}$, $\beta(x) = \log\sqrt{p-x}$, with such a p that $p - x > 0$ for $x \in J$.

Let us remark that the solutions from (iv) are not solutions of the invariance problem, but $\mathcal{A}_{\mu}$ is not a mean. The special case $\lambda = \frac{1}{2}$ was also solved in Głazowska et al. (2002), while the case $\lambda = \mu$ was also solved in Daróczy and Páles (2003a).

2.2 COMPLEMENTARINESS

2.2.1 Complementary means

Given two means M and N, it is very difficult to find a mean P which is (M, N)-invariant, as it was seen in the case of the $\mathcal{AGM}$. Another method was considered to surpass this situation. The idea was taken from Gini (1958), where two means M and N are called **complementary** (with respect to $\mathcal{A}$) if $M + N = 2\mathcal{A}$. We remark that for every mean M, the function $2 \cdot \mathcal{A} - M$ is again a mean. Thus the complementary of every mean M exists and it is denoted by ${}^{c}M$. The most interesting example of mean defined this way is the contraharmonic mean given by

$$\mathcal{C} = {}^{c}\mathcal{H}.$$

But we have also ${}^{c}\mathcal{E} = N_{1/exp}$.

A second notion of this type also considered in Gini (1958) is the following: two means M and N are called **inverses** (with respect to $\mathcal{G}$) if $M \cdot N = \mathcal{G}^2$. Again, for every (non-vanishing) mean M, the expression $\mathcal{G}^2/M$ gives a mean, the inverse of M, which we denote by ${}^iM$. If the mean M is homogeneous, we have

$$ {}^iM(a,b) = \frac{1}{M\left(\frac{1}{b}, \frac{1}{a}\right)} $$

which is used in Borwein and Borwein (1987) as definition of the inverse. For example, we have

$$ {}^i\mathcal{A} = \mathcal{H} \quad \text{and} \quad {}^j\mathcal{I}(a,b) = e\left(\frac{b^a}{a^b}\right)^{\frac{1}{a-b}}. $$

In Toader (1991) and in Matkowski (1999), a generalization of complementariness and of inversion was proposed.

Definition 13. A mean N is called **complementary to M with respect to P** (or P-**complementary** to M) if it verifies

$$ P(M,N) = P. $$

Remark 34. Of course, $\mathcal{A}$-complementary is equivalent to complementary and $\mathcal{G}$-complementary to inverse. Also, N is P-complementary to M if and only if P is (M,N)-invariant. But, in some cases, given the means P and M we are able to determine a mean N (in a family of means) which is P-complementary to M.

Remark 35. The P-complementary of a given mean does not necessarily exist nor is unique. For example, the Π_1-complementary of Π_1 is any mean M, but no mean $M \neq \Pi_1$ has a Π_1-complementary. If a given mean has a unique P-complementary mean N, denote $N = {}^PM$.

Proposition 3. *For every mean M we have*

$$ (i)\ {}^MM = M, $$
$$ (ii)\ {}^M\Pi_1 = \Pi_2, $$
$$ (iii)\ {}^{\Pi_2}M = \Pi_2 $$

and we shall denote also

$$ (iv)\ {}^{\Pi_1}\Pi_1 = M $$

meaning that $\Pi_1(\Pi_1, M) = \Pi_1$ *for every* M.

If P is a symmetric mean then

$$(v)\ {}^{P}\Pi_2 = \Pi_1,$$
$$(vi)\ {}^{P}\wedge = \vee,$$
$$(vii)\ {}^{P}\left({}^{P}M\right) = M.$$

Remark 36. In what follows, we shall call these results as **trivial cases** of complementariness. Of course, we are interested in determining non-trivial cases. The following existence theorem was proved in Matkowski (1999).

Theorem 31. *Let P be a fixed symmetric mean on J which is continuous and strictly isotone. Then every mean M on J has a unique P-complementary mean N on J.*

We want to consider some non-trivial examples. We begin with the case of Greek means, which was studied in Toader (2004). Denote the $\mathcal{F}_j$-complementary of a mean M by ${}^{\mathcal{F}j}M$.

Theorem 32. *We have successively*

$${}^{\mathcal{A}}M = 2\mathcal{A} - M;$$

$${}^{\mathcal{G}}M = \frac{\mathcal{G}^2}{M};$$

$${}^{\mathcal{H}}M = \frac{M \cdot \mathcal{H}}{2M - \mathcal{H}};$$

$${}^{C}M = \frac{1}{2} \cdot \left(C + \sqrt{C^2 + 4MC - 4M^2}\right);$$

$${}^{\mathcal{F}5}M = \begin{cases} \frac{1}{2}\left[\mathcal{F}_5 + \sqrt{\mathcal{F}_5 \cdot (5\mathcal{F}_5 - 4M)}\right], & \text{if } \mathcal{F}_5 \le M \\ \mathcal{F}_5 + M - \frac{M^2}{\mathcal{F}_5}, & \text{if } \mathcal{F}_5 \ge M \end{cases};$$

$${}^{\mathcal{F}6}M = \begin{cases} \mathcal{F}_6 + M - \frac{M^2}{\mathcal{F}_6}, & \text{if } \mathcal{F}_6 \le M \\ \frac{1}{2}\left[\mathcal{F}_6 + \sqrt{\mathcal{F}_6(5\mathcal{F}_6 - 4M)}\right], & \text{if } \mathcal{F}_6 \ge M \end{cases};$$

$${}^{\mathcal{F}7}M = \begin{cases} \frac{1}{2}\left[M + \sqrt{M(4\mathcal{F}_7 - 3M)}\right], & \text{if } \mathcal{F}_7 \le M \\ \frac{1}{2}\left[M + \mathcal{F}_7 + \sqrt{\mathcal{F}_7^2 + 2M\mathcal{F}_7 - 3M^2}\right], & \text{if } \mathcal{F}_7 \ge M \end{cases};$$

$${}^{\mathcal{F}8}M = \begin{cases} 2M - \frac{M^2}{\mathcal{F}_8}, & \text{if } \mathcal{F}_8 \le M \\ \mathcal{F}_8 + \sqrt{\mathcal{F}_8(\mathcal{F}_8 - M)}, & \text{if } \mathcal{F}_8 \ge M \end{cases};$$

$$
{}^{\mathcal{F}9}M = \begin{cases} M - \sqrt{M(M-\mathcal{F}_9)}, & \text{if } \mathcal{F}_9 \le M \\ \frac{M^2}{2M-\mathcal{F}_9}, & \text{if } \mathcal{F}_9 \ge M \end{cases};
$$

$$
{}^{\mathcal{F}10}M = \begin{cases} \frac{1}{2}\left[M + \mathcal{F}_{10} - \sqrt{M^2 + 2M\mathcal{F}_{10} - 3\mathcal{F}_{10}^2}\right], & \text{if } \mathcal{F}_{10} \le M \\ M - \mathcal{F}_{10} + \frac{\mathcal{F}_{10}^2}{M}, & \text{if } \mathcal{F}_{10} \ge M \end{cases}.
$$

Remark 37. If a mean M is not comparable with $\mathcal{F}_i$ (for some $i = 5, \ldots, 10$), then ${}^{\mathcal{F}i}M$ has two expressions, depending on the relation between M and $\mathcal{F}_i$ in the given point. For example, we have:

$$
{}^{\mathcal{F}9}M(a,b) = \begin{cases} M(a,b) - \sqrt{M(a,b)\left[M(a,b) - \mathcal{F}_9(a,b)\right]} \\ \qquad\qquad \text{if } \mathcal{F}_9(a,b) \le M(a,b) \\ \frac{M^2(a,b)}{2M(a,b)-\mathcal{F}_9(a,b)} \qquad \text{if } \mathcal{F}_9(a,b) \ge M(a,b). \end{cases}
$$

The complete list of means which are the complementary of a Greek mean with respect to another is given in Toader and Toader (2004) and Toader and Toader (2004a). As a consequence, the following invariance in the family of Greek means (see Toader and Toader, 2005) was deduced.

Theorem 33. *Among the Greek means we have only the following relations of invariance:*

$$
\mathcal{A}(\mathcal{H}, \mathcal{C}) = \mathcal{A},\ \mathcal{A}(\mathcal{F}_7, \mathcal{F}_9) = \mathcal{A},\ \mathcal{G}(\mathcal{A}, \mathcal{H}) = \mathcal{G},\ \mathcal{G}(\mathcal{F}_8, \mathcal{F}_9) = \mathcal{G},
$$

or their duals.

The **complementariness** with respect to $\mathcal{P}_{m;\lambda}$ for $\lambda \neq 1$ was considered in Costin (2004b). If we denote by ${}^{\mathcal{P}(m;\lambda)}M$ the $P_{m;\lambda}$-complementary of M, we find the expression

$$
{}^{\mathcal{P}(m;\lambda)}M = \left[\frac{(\mathcal{P}_{m;\lambda})^m - \lambda \cdot M^m}{1-\lambda}\right]^{\frac{1}{m}},\ m \neq 0,
$$

while, for $m = 0$ we have

$$
{}^{\mathcal{G}(\lambda)}M = \left(\frac{\mathcal{G}_\lambda}{M^\lambda}\right)^{\frac{1}{1-\lambda}}.
$$

Lemma 12. *The pre-mean* ${}^{\mathcal{P}(m;\lambda)}M$ *is a mean for every mean* M *and each* $m \in \mathbb{R}$ *if and only if* $0 \le \lambda \le \frac{1}{2}$.

Proof. Taking $m > 0$ and $a < b$, we have

$$b > {}^{\mathcal{P}(m;\lambda)}M(a,b) > \left[\frac{\lambda a^m + (1-2\lambda)b^m}{1-\lambda}\right]^{\frac{1}{m}}.$$

The last expression is greater than a if and only if $1 - 2\lambda \geq 0$. Other cases can be treated similarly. □

Remark 38. This complementary mean can exist also for $1/2 < \lambda < 1$, but only for some means. For example, we have

$${}^{\mathcal{G}(\lambda)}\mathcal{G}_{\mu} = \mathcal{G}_{\frac{\lambda(1-\mu)}{1-\lambda}} \quad \text{for } 0 < \lambda \leq \frac{1}{2-\mu}. \tag{2.20}$$

We shall refer also to the following special cases

$${}^{\mathcal{G}(\lambda)}\mathcal{G}_{\frac{3\lambda-1}{2\lambda}} = \mathcal{G} \tag{2.21}$$

$${}^{\mathcal{G}(\lambda)}\mathcal{G} = \mathcal{G}_{\frac{\lambda}{2(1-\lambda)}} \tag{2.22}$$

$${}^{\mathcal{G}(1/3)}\Pi_2 = \mathcal{G} \tag{2.23}$$

$${}^{\mathcal{G}(2/3)}\mathcal{G} = \Pi_1 \tag{2.24}$$

$${}^{\mathcal{G}(\lambda)}\mathcal{G}_{\frac{2\lambda-1}{\lambda}} = \Pi_1$$

$${}^{\mathcal{G}}\mathcal{G}_{\mu} = \mathcal{G}_{1-\mu}$$

Remark 39. The complementary of a mean with respect to an arbitrary mean, like $\mathcal{I}$, is generally difficult to determine. To find some necessary conditions on a mean to be the complementary of a given mean with respect to another, series expansions of means will be used in what follows.

Let us give a comparability result for the complementaries with respect to a given mean.

Theorem 34. *If the isotonic mean P is (M, N)-invariant and (M', N')-invariant, then $M \leq M'$ implies $N \geq N'$.*

Proof. We have

$$P(M, N) = P = P(M', N')$$

and

$$P(M, N) \leq P(M', N),$$

thus

$$P\left(M', N'\right) \leq P\left(M', N\right),$$

implying $N' \leq N$. □

Corollary 10. *If* $M \leq N$ *then* ${}^{\mathcal{G}}M \geq {}^{\mathcal{G}}N$.

Example 3. The sequence of inequalities from (2.10) can be augmented as:

$$\mathcal{A} > \mathcal{I} > \mathcal{L} > \mathcal{G} = {}^{\mathcal{G}}\mathcal{G} > {}^{\mathcal{G}}\mathcal{L} > {}^{\mathcal{G}}\mathcal{I} > {}^{\mathcal{G}}\mathcal{A} = \mathcal{H}.$$

Remark 40. The mean $\mathcal{C}$ is not isotonic and ${}^{\mathcal{C}}\mathcal{A}$ is not comparable with ${}^{\mathcal{C}}\mathcal{G}$.

2.2.2 Algebraic and topological structures on some set of means

In Farhi (2010), some structures on the set M_D of strict symmetric means defined on a nonempty symmetric domain $D \subset \mathbb{R}^2$ are presented. First of all on M_D a structure of abelian group is defined. For this, the set A_D of asymmetric maps on D is considered, that is the maps $f : D \to \mathbb{R}_+$ satisfying:

$$f(y, x) = -f(x, y), \ \forall\, (x, y) \in D.$$

It is clear that $(A_D, +)$ is an abelian group with the neutral element the null map. It is easy to prove that the map $\varphi : M_D \to A_D$ given by

$$\varphi(M)(x, y) = \begin{cases} \log \dfrac{M(x, y) - x}{y - M(x, y)}, & \text{if} \quad x \neq y \\ 0, & \text{if} \quad x = y \end{cases}$$

is well defined. Moreover, φ is a bijection and its inverse map φ^{-1} for $f \in A_D$ is given by:

$$\varphi^{-1}(f)(x, y) = M(x, y) = \frac{x + ye^{f(x,y)}}{e^{f(x,y)} + 1}, \ (x, y) \in D.$$

Of course $\varphi(M) = f$. Through φ, the abelian structure $(A_D, +)$ can be transported on M_D. For this, the following internal composition law $*$ is defined on M_D:

$$M * N = \varphi^{-1}(\varphi(M) + \varphi(N)).$$

As $\varphi^{-1}(0) = \mathcal{A}$, we deduce that the arithmetic mean is the neutral element of the group $(M_D, *)$.

Given a fixed $P \in (M_D, *)$, the symmetry with respect to P in the group is defined by:

$$S_P(M) = N \text{ if and only if } M * N = P * P.$$

It is proved that

$$N(x, y) = \frac{x(M-x)(P-y)^2 - y(P-x)^2(M-y)}{(M-x)(P-y)^2 - (P-x)^2(M-y)},$$

where, for simplicity, it was denoted M instead of $M(x, y)$ and P instead of $P(x, y)$. For instance:

$$S_{\mathcal{A}}(M) = 2\mathcal{A} - M = {}^{\mathcal{A}}M,$$
$$S_{\mathcal{G}}(M) = \frac{\mathcal{G}^2}{M} = {}^{\mathcal{G}}M,$$
$$S_{\mathcal{H}}(M) = \frac{\mathcal{H}M}{2M - \mathcal{H}} = {}^{\mathcal{H}}M.$$

It is an open question to determine all the means P for which

$${}^{P}M = S_P(M) \text{ for each } M \in M_D$$

in the case $D = (0, +\infty)^2$.

In the same paper a metric space structure on M_D is constructed. For $M, N \in M_D$, define

$$d(M, N) = \sup_{(x,y)\in D,\ x\neq y} \left| \frac{M(x, y) - N(x, y)}{x - y} \right|. \tag{2.25}$$

It is proved that d is a metric on M_D and the metric space (M_D, d) is identical to the closed ball of center $\mathcal{A}$ and radius $1/2$. If we denote $\varphi(M) = f$ and $\varphi(N) = g$, then

$$d(M, N) = \sup_{(x,y)\in D} \left(\frac{1}{e^{f(x,y)} + 1} - \frac{1}{e^{g(x,y)} + 1} \right).$$

If $D = J^2$, where J is an interval of $\mathbb{R}$ and $d(M, N) < 1$, it is proved that there exists an unique mean P which is (M, N)-invariant. The result is not necessarily valid if $d(M, N) = 1$.

There are other papers that deal with algebraic and topological structures on sets of means, for example Matkowski (2002a).

2.2.3 More about pre-means

Some of the definitions related to means were generalized in Toader and Toader (2007) to allow the study of relations between pre-means. Let D be a set in $\mathbb{R}_+^2$ and M be a real function defined on D.

Definition 14. The function M is called a **mean** on D if it has the property

$$a \wedge b \le M(a,b) \le a \vee b, \ \forall (a,b) \in D.$$

In the special case $D = J^2$, where $J \subset \mathbb{R}_+$ is an interval, the function M is called **mean** on J.

Definition 15. The function M is called a **pre-mean** on D if it has the property

$$M(a,a) = a, \ \forall (a,a) \in D.$$

A function M (not necessarily a mean) can have some special properties.

Definition 16. i) The function M is **symmetric** on D if $(a,b) \in D$ implies $(b,a) \in D$ and

$$M(a,b) = M(b,a), \ \forall (a,b) \in D.$$

ii) The function M is **homogeneous** (of degree one) on D if there exists a neighborhood V of 1 such that $t \in V$ and $(a,b) \in D$ implies $(ta,tb) \in D$ and

$$M(ta,tb) = tM(a,b).$$

iii) The function M is **strict at the left** (respectively **strict at the right**) on D if for $(a,b) \in D$

$$M(a,b) = a \text{ (respectively } M(a,b) = b\text{), implies } a = b.$$

iv) The function M is **strict** if it is strict at the left and strict at the right.

Remark 41. Each mean is reflexive, thus it is a pre-mean. Conversely, each pre-mean on D is a mean on $D \cap \Delta$, where

$$\Delta = \{(a,a); a \ge 0\}.$$

The question is if the set $D \cap \Delta$ can be extended. The answer is generally negative. Take for instance the pre-mean $S_{r,s;\lambda}$ for $\lambda \notin [0,1]$. Even though it is defined on a larger set like

$$\left(\frac{\lambda}{\lambda-1}\right)^{1/s} \le \frac{b}{a} \le \left(\frac{\lambda}{\lambda-1}\right)^{1/r}, \text{ for } \lambda > 1, r > s > 0,$$

it is a mean only on Δ. However, the above question may have also a positive answer. For example, a special result from Abramovich and Pečarić (2000) was generalized in Toader (2002a) as follows.

Theorem 35. *If M is a differentiable homogeneous pre-mean on $\mathbb{R}_+^2$ such that*

$$0 < M_b(1,1) = \frac{\partial}{\partial b} M(1,1) < 1,$$

then there exists the constants $T' < 1 < T''$ such that M is a mean on

$$D = \{(a,b) \in \mathbb{R}_+^2; T'a \le b \le T''a\}.$$

The previous result was strengthened in Toader and Toader (2007), by renouncing the hypothesis of homogeneity of the pre-mean M.

Theorem 36. *If M is a differentiable pre-mean on the open set D such that*

$$0 < M_b(a,a) < 1, \forall (a,a) \in D,$$

then for each $(a,a) \in D$ there exist the constants $T'_a < 1 < T''_a$ such that

$$ta \le M(a,ta) \le a; T'_a \le t \le 1$$

and

$$a \le M(a,ta) \le ta; 1 \le t \le T''_a.$$

Example 4. For $M = \mathcal{A}_\lambda^2/\mathcal{G}$ as $M_b(1,1) = (3-4\lambda)/2$, the previous result is valid for M if $\lambda \in (0.25, 0.75)$. A table with the interval (T', T'') for some values of λ is the following:

λ	T'	T''
0.25	0.004...	1.
0.3	0.008...	1.671...
0.5	0.087...	11.444...
0.7	0.598...	113.832...
0.75	1.0	243.776...

For $\lambda \notin [0.25, 0.75]$ one gets $T' = T'' = 1$.

Remark 42. A similar result can be proved in the case

$$0 < M_a(b,b) < 1, \forall (b,b) \in D.$$

If the partial derivatives do not belong to the interval $(0,1)$, the result can be false.

Example 5. For $M = \mathcal{S}_{r,s;\lambda}$, we have $M_b(a,a) = 1 - \lambda$. As we remarked, for $\lambda \notin [0, 1]$ the pre-Gini mean is a mean only on Δ.

2.2.4 Complementary pre-means

Given three functions M, N, and P on D, consider the subset $D' \subseteq D$ such that $(M(a,b), N(a,b)) \in D$, $\forall (a,b) \in D'$. The **composition** $P(M,N)$ can be defined on D' by

$$P(M,N)(a,b) = P(M(a,b), N(a,b)), \ \forall (a,b) \in D'.$$

If M, N, and P are means then $D' = D$.

Definition 17. A function N is called **complementary to** M **with respect to** P (or P-**complementary** to M) if it verifies

$$P(M,N) = P \text{ on } D'.$$

Remark 43. In the same circumstances, the function P is called (M,N)-**invariant**. If M has a unique P-complementary N, denote it by $N = {}^{P}M$.

Remark 44. If P and M are means, the P-complementary of M is generally not a mean, as we saw in some examples. For pre-means we get the following result.

Theorem 37. *If P and M are pre-means and P is strict at the left, then the P-complementary of M is a pre-mean N.*

Proof. We have

$$P(M(a,a), N(a,a)) = P(a,a), \ \forall (a,a) \in D,$$

thus

$$P(a, N(a,a)) = a, \ \forall (a,a) \in D$$

and as P is strict at the left, we get $N(a,a) = a$, $\forall (a,a) \in D$. □

The result cannot be improved for means, thus we have only the following:

Corollary 11. *If P and M are means and P is strict at the left, then the P-complementary of M is a pre-mean N.*

2.2.5 Partial derivatives of pre-means

Regarding the first order partial derivatives of pre-means defined on J, the following results were proved in Toader (2002a).

Theorem 38. *If M is a differentiable pre-mean then*

$$M_a(c,c) + M_b(c,c) = 1, \forall c \in J. \tag{2.26}$$

Theorem 39. *If M is a differentiable mean then*

$$M_a(c,c) \geq 0, \forall c \in J.$$

Theorem 40. *If M is a symmetric differentiable pre-mean then*

$$M_a(c,c) = M_b(c,c) = 1/2, \forall c \in J. \tag{2.27}$$

Example 6. This property is valid for the Greek means

$$\mathcal{A}, \mathcal{G}, \mathcal{H}, \mathcal{C}, \mathcal{F}_5 \text{ and } \mathcal{F}_6,$$

but the other are not differentiable.

Remark 45. For a mean, the relation (2.27) was proved in Foster and Phillips (1984).

In Toader (2002a), also some results on second order partial derivatives of pre-means are given.

Theorem 41. *If M is a twice differentiable pre-mean then*

$$M_{aa}(c,c) + 2M_{ab}(c,c) + M_{bb}(c,c) = 0, \forall c \in J. \tag{2.28}$$

Corollary 12. *If M is a symmetric pre-mean then*

$$M_{ab}(c,c) = -M_{aa}(c,c), \forall c \in J. \tag{2.29}$$

Most of the "usual" symmetric means have also the property

$$M_{aa}(c,c) = \frac{\alpha}{c}, \ \alpha \in \mathbb{R}, \forall c \in J. \tag{2.30}$$

By direct (but difficult) computation, the following values of α in (2.30) were found:

M	$\mathcal{D}_{r,s}$	$\mathcal{S}_{r,s}$
α	$\frac{r+s-3}{12}$	$\frac{r+s-1}{4}$

.

Of course, here are included the values of $\mathcal{L}_r, \mathcal{I}_r$ and $\mathcal{P}_r = \mathcal{S}_{r,0} = \mathcal{D}_{2r,r}$. For the first six Greek means, which are differentiable, we have the values

M	$\mathcal{A}$	$\mathcal{G}$	$\mathcal{H}$	$\mathcal{C}$	$\mathcal{F}_5$	$\mathcal{F}_6$
α	0	$-\frac{1}{4}$	$-\frac{1}{2}$	$\frac{1}{2}$	$\frac{1}{4}$	$\frac{1}{4}$

.

But not all the symmetric means have this property. For example, for the mean $\mathcal{E}$ given by (2.8), we obtain

$$\mathcal{E}_{aa}(c,c) = \frac{1}{6}, \forall c \in J,$$

thus it is not of the type (2.30).

On the other hand a non-symmetric mean can also have the above property. For example, in the case of $\mathcal{P}_{n,\lambda}$ we have (2.30) with

$$\alpha = \lambda(1-\lambda)(n-1).$$

For higher order partial derivatives, by mathematical induction, the following result was proved.

Theorem 42. *If the pre-mean M has a n-th order differential, then*

$$\sum_{j=0}^{k} \binom{k}{j} \frac{\partial^k}{\partial^j \partial^{k-j}} M(c,c) = 0, \ 1 \le k \le n, \forall c \in J.$$

Corollary 13. *If the symmetric pre-mean M has a third order differential, then*

$$M_{a^3}(c,c) = M_{b^3}(c,c) = -3M_{a^2b}(c,c) = -3M_{a^2b}(c,c), \forall c \in J. \quad (2.31)$$

2.2.6 Series expansion of means

For the study of some problems related to means, the power series expansion was used in Lehmer (1971). Let M be a symmetric and homogeneous mean. Without loss of generality we may assume that M acts on the positive numbers $a \ge b$ and

$$M(a,b) = aM(1,b/a) = aM(1,1-t),$$

where

$$0 \le t = 1 - b/a < 1.$$

For many problems it suffices to consider only the normalized function $M(1, 1-t)$ even if the mean M is not symmetric nor homogeneous. We shall give explicit Taylor series coefficients of the normalized function for some means. To avoid the complication of the presentation, we shall call them series expansions of the corresponding means.

Example 7. In the case of Greek means we need only the binomial series (1.29) for obtaining their expansions. Using the denotation of (1.30), the following results were given in Toader and Toader (2003a):

$$\mathcal{A}(1,1-t)=1-\frac{t}{2};$$

$$\mathcal{G}(1,1-t)=(1-t)^{1/2}=\sum_{n=0}^{\infty}(-1)^n\left(\frac{1}{2}\right)_n\frac{t^n}{n!}=1-\frac{t}{2}-\frac{t^2}{8}-\frac{t^3}{16}-\frac{5t^4}{128}+\cdots;$$

$$\mathcal{H}(1,1-t)=\frac{2(1-t)}{2-t}=2-\left(1-\frac{t}{2}\right)^{-1}=1-\sum_{n=1}^{\infty}\left(\frac{t}{2}\right)^n$$

$$=1-\frac{t}{2}-\frac{t^2}{4}-\frac{t^3}{8}-\frac{t^4}{16}+\cdots;$$

$$\mathcal{C}(1,1-t)=\frac{1+(1-t)^2}{2-t}=\left(1-\frac{t}{2}\right)^{-1}-t=1-\frac{t}{2}+\sum_{n=2}^{\infty}\left(\frac{t}{2}\right)^n$$

$$=1-\frac{t}{2}+\frac{t^2}{4}+\frac{t^3}{8}+\frac{t^4}{16}+\cdots;$$

$$\mathcal{F}_5(1,1-t)=\frac{1}{2}\left[t+\sqrt{t^2+4(1-t)^2}\right]=\frac{t}{2}+\left(1-2t+\frac{5t^2}{4}\right)^{1/2}$$

$$=1-\frac{t}{2}+\frac{5t^2}{8}+\sum_{n=2}^{\infty}\left(\frac{1}{2}\right)_n\frac{1}{n!}\left(\frac{5t^2}{4}-2t\right)^n=1-\frac{t}{2}+\frac{t^2}{8}+\frac{t^3}{8}+\frac{95t^4}{128}+\cdots;$$

$$\mathcal{F}_6(1,1-t)=\frac{1}{2}\left(\sqrt{t^2+4}-t\right)=\left[1+\left(\frac{t}{2}\right)^2\right]^{1/2}-\frac{t}{2}$$

$$=1-\frac{t}{2}+\sum_{n=1}^{\infty}\left(\frac{1}{2}\right)_n\frac{1}{n!}\left(\frac{t}{2}\right)^{2n}=1-\frac{t}{2}+\frac{t^2}{8}-\frac{t^4}{128}+\cdots;$$

$$\mathcal{F}_7(1,1-t)=1-t+t^2;$$

$$\mathcal{F}_8(1,1-t)=\frac{1}{1+t}=\sum_{n=0}^{\infty}(-1)^n t^n=1-t+t^2-t^3+t^4+\cdots;$$

$$\mathcal{F}_9(1,1-t)=1-t^2;$$

$$\mathcal{F}_{10}(1,1-t)=\frac{1}{2}\left[1-t+\left(1+2t-3t^2\right)^{1/2}\right]$$

$$= 1 - \frac{3t^2}{4} + \frac{1}{2}\sum_{n=2}^{\infty}\left(\frac{1}{2}\right)_n \frac{1}{n!}\left(2t - 3t^2\right)^n = 1 - t^2 + t^3 - \frac{5t^4}{4} + \cdots .$$

We remark that three of these normalized functions are polynomials.

Remark 46. Similarly we get

$$\mathcal{G}_\lambda(1, 1-t) = 1 - (1-\lambda)t - \frac{\lambda(1-\lambda)}{2}t^2 - \frac{\lambda\left(1-\lambda^2\right)}{6}t^3$$
$$- \frac{\lambda\left(1-\lambda^2\right)(2+\lambda)}{24}t^4 + \cdots .$$

Remark 47. The determination of all the coefficients, for some means is impossible. In these cases, a recurrence relation for the coefficients will be very useful. It gives a way to calculate as many coefficients as desired. To obtain it, we can apply an old **formula of Euler** presented, together with its history, in Gould (1974).

Theorem 43. *If the function f has the Taylor series expansion*

$$f(x) = \sum_{n=0}^{\infty} a_n \cdot x^n,$$

p is a real number, and

$$[f(x)]^p = \sum_{n=0}^{\infty} b_n \cdot x^n,$$

then we have the recurrence relation

$$\sum_{k=0}^{n} [k(p+1) - n] \cdot a_k \cdot b_{n-k} = 0, \; n \geq 0. \tag{2.32}$$

Proof. We use the differentiation of formal power series which is also justified in Gould (1974). Starting from the identity

$$D_x[f(x)]^{p+1} = (p+1)[f(x)]^p D_x f(x) = D_x\{[f(x)]^p \cdot f(x)\},$$

we have

$$(p+1)\left(\sum_{n=0}^{\infty} b_n x^n\right)\sum_{m=1}^{\infty} m a_m x^{m-1} = D_x\left[\left(\sum_{n=0}^{\infty} b_n x^n\right)\sum_{m=0}^{\infty} a_m x^m\right].$$

Multiplying the series, we get

$$(p+1)\sum_{n=1}^{\infty}\left(x^{n-1}\sum_{k=0}^{n}k\cdot a_k\cdot b_{n-k}\right)=D_x\left[\sum_{n=0}^{\infty}\left(x^{n}\sum_{k=0}^{n}a_k\cdot b_{n-k}\right)\right],$$

thus

$$\sum_{n=1}^{\infty}\left(x^{n-1}\sum_{k=0}^{n}k(p+1)a_kb_{n-k}\right)=\sum_{n=1}^{\infty}\left(nx^{n-1}\sum_{k=0}^{n}a_kb_{n-k}\right).$$

Equating coefficients, we have proved (2.32). □

Corollary 14. *If $a_0 \neq 0$ then*

$$b_0 = a_0^p$$

and

$$b_n=\frac{1}{n\cdot a_0}\sum_{k=1}^{n}[k(p+1)-n]\cdot a_k\cdot b_{n-k}\ ,\ n\geq 1. \tag{2.33}$$

Corollary 15. *If the function f has the Taylor series expansion*

$$f(x)=\sum_{n=0}^{\infty}a_n\cdot x^n,$$

and p is a real number, then the Taylor series of $\left[f(x)\right]^p$ has the following first coefficients

$$\begin{aligned}\left[f(x)\right]^p &= a_0^p+pa_0^{p-1}a_1\cdot x+pa_0^{p-2}\left[(p-1)\frac{a_1^2}{2!}+a_0a_2\right]\cdot x^2\\ &+pa_0^{p-3}\left[(p-1)(p-2)\frac{a_1^3}{3!}+(p-1)a_0a_1a_2+a_0^2a_3\right]\cdot x^3\\ &+pa_0^{p-4}\left[(p-1)(p-2)(p-3)\frac{a_1^4}{4!}+(p-1)(p-2)a_0\frac{a_1^2}{2!}a_2\right.\\ &\left.+(p-1)\left(a_1a_3+\frac{a_2^2}{2}\right)a_0^2+a_0^3a_4\right]\cdot x^4+pa_0^{p-5}\left[(p-1)(p-2)\right.\\ &\cdot(p-3)(p-4)\frac{a_1^5}{5!}+(p-1)(p-2)(p-3)a_0\frac{a_1^3}{3!}a_2+(p-1)(p-2)\\ &\left.\cdot\frac{a_0^2a_1}{2}\left(a_2^2+a_1a_3\right)+(p-1)a_0^3(a_2a_3+a_1a_4)+a_0^4a_5\right]\cdot x^5+pa_0^{p-6}\end{aligned}$$

$$\cdot\left[(p-1)(p-2)(p-3)(p-4)(p-5)\frac{a_1^6}{6!}\right.$$

$$+(p-1)(p-2)(p-3)(p-4)\,a_0\frac{a_1^4}{4!}a_2$$

$$+(p-1)(p-2)(p-3)\,a_0^2\frac{a_1^2}{4}\left(a_2^2+\frac{2a_1a_3}{3}\right)+(p-1)(p-2)\frac{a_0^3}{6}$$

$$\left.\cdot\left(3a_1^2a_4+6a_1a_2a_3+a_2^3\right)+(p-1)\,a_0^4\left(a_1a_5+a_2a_4+\frac{a_3^2}{2}\right)+a_0^5a_6\right]\cdot x^6$$

$$+\cdots.$$

Lemma 13. *Let the function*

$$f_{p,q;\lambda}(x)=\left[\lambda+(1-x)^p\right]^{\frac{1}{q}},\ \lambda\neq-1.$$

Its Taylor series coefficients

$$f_{p,q;\lambda}(x)=\sum_{n=0}^{\infty}x^n a_n(p,q,\lambda),$$

verify the recurrence relation

$$a_n(p,q,\lambda)=\frac{1}{nq(1+\lambda)}\sum_{k=0}^{n-1}(-1)^{n-k+1}\binom{p}{n-k}[k(q+1)-n]\cdot a_k(p,q,\lambda)$$

for $n\geq 1$, with

$$a_0(p,q,\lambda)=(1+\lambda)^{\frac{1}{q}}.$$

Proof. As

$$\left[f_{p,q;\lambda}(x)\right]^q=\lambda+(1-x)^p=\sum_{n=0}^{\infty}x^n b_n(p,\lambda),$$

we get

$$b_0(p,\lambda)=\lambda+1,\ b_n(p,\lambda)=(-1)^n\binom{p}{n},\ n\geq 1.$$

The recurrence relation (2.32) is now

$$\sum_{k=0}^{n}[k(q+1)-n]\cdot a_k(p,q,\lambda)\cdot b_{n-k}(p,\lambda)=0,\ n\geq 0,$$

which implies

$$nq(1+\lambda)\cdot a_n(p,q,\lambda)$$
$$+\sum_{k=0}^{n-1}(-1)^{n-k}\binom{p}{n-k}[k(q+1)-n]\cdot a_k(p,q,\lambda)=0,$$

$n \geq 1$. It gives the recurrence relation. □

Corollary 16. *The first terms of the Taylor series of the function* $f_{p,q;\lambda}$ *are given by*

$$
\begin{aligned}
f_{p,q;\lambda}(x) &= (1+\lambda)^{\frac{1}{q}}\cdot\left\{1-\frac{p\cdot x}{q\cdot(1+\lambda)}\right.\\
&+\left[q\,(p-1)\,(1+\lambda)-p\,(q-1)\right]\cdot\frac{p\cdot x^2}{2\cdot q^2\cdot(1+\lambda)^2}\\
&-[q^2\cdot(p-1)\,(p-2)\,(1+\lambda)^2-3qp\cdot(p-1)\,(q-1)\,(1+\lambda)\\
&+p^2(q-1)(2q-1)]\cdot\frac{p\cdot x^3}{6\cdot q^3\cdot(1+\lambda)^3}+[q^3\cdot(p-1)\,(p-2)\,(p-3)\,(1+\lambda)^3\\
&-pq^2\,(p-1)\,(7p-11)\,(q-1)\,(1+\lambda)^2+6qp^2\,(p-1)\,(q-1)(2q-1)\,(1+\lambda)\\
&-p^3\cdot(q-1)(2q-1)(3q-1)]\cdot\frac{p\cdot x^4}{24\cdot q^4\cdot(1+\lambda)^4}\\
&-[q^4\cdot(p-1)\,(p-2)\,(p-3)\,(p-4)\,(1+\lambda)^4\\
&-5pq^3\cdot(p-1)\,(p-2)\,(3p-5)\,(q-1)\\
&\cdot(1+\lambda)^3+5p^2q^2\cdot(p-1)\,(5p-7)\,(q-1)(2q-1)\cdot(1+\lambda)^2\\
&-10p^3q\cdot(p-1)\,(q-1)(2q-1)(3q-1)\cdot(1+\lambda)\\
&+p^4\cdot(q-1)(2q-1)(3q-1)(4q-1)]\cdot\frac{p\cdot x^5}{120q^5\cdot(1+\lambda)^5}\\
&+[q^5\cdot(p-1)\,(p-2)\,(p-3)\,(p-4)\,(p-5)\,(1+\lambda)^5\\
&-pq^4\cdot(p-1)\,(p-2)\left(31p^2-132p+137\right)(q-1)\,(1+\lambda)^4\\
&+15p^2q^3\cdot(p-1)\,(2p-3)\,(3p-5)\,(q-1)(2q-1)\cdot(1+\lambda)^3\\
&-5p^3q^2\,(p-1)\,(13p-17)\,(q-1)(2q-1)(3q-1)\,(1+\lambda)^2\\
&+15p^4q\cdot(p-1)\,(q-1)(2q-1)(3q-1)(4q-1)\cdot(1+\lambda)\\
&\left.-p^5(q-1)(2q-1)(3q-1)(4q-1)(5q-1)]\cdot\frac{p\cdot x^6}{720q^6\cdot(1+\lambda)^6}+\cdots\right\}.
\end{aligned}
$$

The following consequence was proved in Costin (2003).

Corollary 17. *For the weighted power mean $\mathcal{P}_{q;t}$ we have*

$$\mathcal{P}_{q;t}(1,1-x)=\left[t+(1-t)(1-x)^q\right]^{\frac{1}{q}}=(1-t)^{\frac{1}{q}}\cdot f_{q,q,t/(1-t)}(x)$$
$$=(1-t)^{\frac{1}{q}}\cdot\sum_{n=0}^{\infty}x^n a_n\left(q,q,\frac{t}{1-t}\right)$$

and we can determine the first terms of its Taylor series

$$\begin{aligned}\mathcal{P}_{q;t}(1,1-x)&=1-(1-t)\cdot x+t(1-t)(q-1)\cdot\frac{x^2}{2!}-t(1-t)(q-1)\\&\cdot\left[t\,(2q-1)-(q+1)\right]\cdot\frac{x^3}{3!}+t(1-t)(q-1)[t^2\,(3q-1)\,(2q-1)\\&-3t\,(2q-1)\,(q+1)+(q+1)\,(q+2)]\cdot\frac{x^4}{4!}\\&-t(1-t)(q-1)[t^3\,(4q-1)\,(3q-1)\,(2q-1)\\&-6t^2\,(3q-1)\,(2q-1)\,(q+1)\\&+t\,(2q-1)\,(q+1)\,(7q+11)-(q+1)\,(q+2)\,(q+3)]\cdot\frac{x^5}{5!}\\&+t(1-t)(q-1)\cdot[t^4\,(5q-1)\,(4q-1)\,(3q-1)\,(2q-1)\\&-10t^3\,(4q-1)\,(3q-1)\,(2q-1)\,(q+1)\\&+5t^2\,(3q-1)\,(2q-1)\,(q+1)\,(5q+7)\\&-5t\,(2q-1)\,(q+1)\,(q+2)\,(3q+5)\\&+(q+1)\,(q+2)\,(q+3)\,(q+4)]\cdot\frac{x^6}{6!}+\cdots.\end{aligned}$$

Corollary 18. *For $t=1/2$ we get*

$$\begin{aligned}\mathcal{P}_q(1,1-x)&=1-\frac{x}{2}+(q-1)\frac{x^2}{8}+(q-1)\frac{x^3}{16}-(q-1)(q-3)(2q+5)\frac{x^4}{384}\\&-(q-1)(2q^2-q-7)\frac{x^5}{256}+(q-1)\,(q-5)\,(16q^3+66q^2-79q-189)\\&\cdot\frac{x^6}{46080}+\ldots\end{aligned}$$

Its first part was given in Lehmer (1971).

In Costin and Toader (2003), the following was proved.

Corollary 19. *For the weighted Gini mean $\mathcal{S}_{p,p-r;t}$, with $r \neq 0$, $t \in (0,1)$, we have*

$$\mathcal{S}_{p,p-r;t}(1,1-x) = \left[\frac{\lambda + (1-x)^p}{\lambda + (1-x)^{p-r}}\right]^{\frac{1}{r}} = f_{p,r,\lambda}(x) \cdot f_{p-r,-r,\lambda}(x),$$

where

$$\lambda = \frac{t}{1-t}.$$

Thus its first terms are

$$\begin{aligned}
&\mathcal{S}_{p,p-r;t}(1,1-x) = 1 - (1-t)\cdot x + t(1-t)(2p-r-1)\cdot\frac{x^2}{2!} - t \\
&\cdot(1-t)[t(6p^2 - 6p(r+1) + (r+1)(2r+1)) - 3p(p-r) - (r-1)(r+1)] \\
&\cdot\frac{x^3}{3!} - t(1-t)\cdot[t^2(-24p^3 + 36p^2(r+1) - 12p(r+1)(2r+1) \\
&+(r+1)(2r+1)(3r+1)) + t(24p^3 - 12p^2(3r+1) + 12p(r+1)(2r-1) \\
&-3(r+1)(2r+1)(r-1)) - 4p^3 + 6p^2(r-1) - 2p(2r^2 - 3r - 1) \\
&+(r-2)(r-1)(r+1)]\cdot\frac{x^4}{4!} - t(1-t)\cdot[t^3(120p^4 - 240p^3(r+1) \\
&+120p^2(r+1)(2r+1) - 20p(r+1)(2r+1)(3r+1) \\
&+(r+1)(2r+1)(3r+1)(4r+1)) + t^2(-180p^4 + 180p^3(2r+1) \\
&-90p^2(r+1)(4r-1) + 30p(r+1)(2r+1)(3r-2) \\
&-6(r-1)(r+1)(2r+1)(3r+1)) + t(70p^4 - 20p^3(7r-2) \\
&+10p^2(14r^2 - 6r - 9) - 10p(r+1)(7r^2 - 12r + 3) \\
&+(r-1)(2r+1)(7r-11)(r+1)) - 5p^4 + 10p^3(r-2) - 5p^2(2r^2 - 6r + 3) \\
&+5p(r-2)(r^2 - 2r - 1) - (r+1)(r-1)(r-2)(r-3)]\cdot\frac{x^5}{5!} - t \\
&\cdot(1-t)[t^4(-720p^5 + 1800p^4(r+1) - 1200p^3(r+1)(2r+1) \\
&+300p^2(r+1)(2r+1)(3r+1) - 30p(r+1)(2r+1)(3r+1)(4r+1) \\
&+(r+1)(2r+1)(3r+1)(4r+1)(5r+1)) + t^3(1440p^5 - 720p^4(5r+3) \\
&+480p^3(r+1)(10r-1) - 120p^2(r+1)(2r+1)(15r-7) \\
&+60p(r+1)(2r+1)(3r+1)(4r-3) - 10(r+1)(2r+1)(3r+1)(4r+1) \\
&\cdot(r-1)) + t^2(-900p^5 + 90p^4(25r+1) - 60p^3(50r^2 + 3r - 26) \\
&+90p^2(r+1)(25r^2 - 27r + 1) - 30p(r+1)(2r+1)(15r^2 - 27r + 10)
\end{aligned}$$

$$+5(r+1)(2r+1)(3r+1)(r-1)(5r-7))$$
$$+t(180p^5-30p^4(15r-13)+60p^3(10r^2-13r-2)$$
$$-30p^2(15r^3-68r^2-6r+17)+30p(r+1)(2r-3)(3r^2-6r+1)$$
$$-5(r+1)(2r+1)(r-1)(r-2)(3r-5))-6p^5+15p^4(r-3)$$
$$-10p^3(r-2)(2r-5)+15p^2(r-3)(r^2-3r+1)$$
$$-p(6r^4-45r^3+100r^2-45r-52)+(r+1)(r-1)(r-2)(r-3)]$$
$$\cdot\frac{x^6}{720}+\cdots$$

Taking $r=1$ we get:

Corollary 20. *For the weighted Lehmer mean $C_{p;t}$ we have*

$$C_{p;t}(1,1-x)=1-(1-t)\cdot x+t(1-t)(p-1)\cdot x^2$$
$$-(p-1)(2t(p-1)-p)\cdot\frac{x^3}{2}$$
$$+t(1-t)(p-1)[6t^2(p-1)^2-6tp(p-1)+p(p+1)]$$
$$\cdot\frac{x^4}{6}-t(1-t)(p-1)\cdot[24t^3(p-1)^3-36t^2p(p-1)^2+2pt(p-1)(7p+4)$$
$$-p(p+1)(p+2)]\cdot\frac{x^5}{24}+t(1-t)(p-1)[120t^4(p-1)^4-240t^3p(p-1)^3$$
$$+30t^2p(p-1)^2(5p+2)-10tp(p-1)(p+1)(3p+2)$$
$$+p(p+1)(p+2)(p+3)]\cdot\frac{x^6}{120}+\cdots.$$

Corollary 21. *For the Lehmer mean C_p we have*

$$C_p(1,1-x)=1-\frac{1}{2}\cdot x+\frac{p-1}{4}\cdot x^2+\frac{p-1}{8}\cdot x^3$$
$$-\frac{(p-1)(p+1)(p-3)}{48}\cdot x^4-\frac{(p-1)\left(p^2-2p-1\right)}{32}\cdot x^5$$
$$+\frac{(p-1)(p+3)(p-5)\left(2p^2-4p-1\right)}{960}\cdot x^6+\cdots.$$

The first terms were given in Gould and Mays (1984).

Corollary 22. *For the Gini means $\mathcal{S}_{p,q}$ we have*

$$\mathcal{S}_{p,q}(1,1-x)=1-\frac{1}{2}\cdot x+\frac{p+q-1}{8}\cdot x^2+\frac{p+q-1}{16}\cdot x^3$$

$$-\left[2(p^3+p^2q+pq^2+q^3)-3(p+q)^2-14(p+q)+15\right]\cdot\frac{x^4}{384}$$
$$-\left[2(p^3+p^2q+pq^2+q^3)-3(p+q)^2-6(p+q)+7\right]\cdot\frac{x^5}{256}$$
$$+[16\left(p^5+p^4q+p^3q^2+p^2q^3+pq^4+q^5\right)-30(p^4+2p^3q+2p^2q^2$$
$$+2pq^3+q^4)-395(p^3+q^3)-365pq(p+q)+615(p+q)^2$$
$$+739(p+q)-945]\cdot\frac{x^6}{46080}+\cdots.$$

Remark 48. In the case of Stolarsky means, we have

$$\mathcal{D}_{r,s}(1,1-x)=\left(\frac{r}{s}\cdot\frac{1-(1-x)^s}{1-(1-x)^r}\right)^{\frac{1}{s-r}}$$

and if we write

$$\mathcal{D}_{r,s}(1,1-x)=g_{s,r}(x)\cdot g_{r,s}(x),$$

with

$$g_{s,r}(x)=\left(\frac{1-(1-x)^s}{s}\right)^{\frac{1}{s-r}},$$

we cannot use the above method for getting the power series expansion, because $g_{s,r}(0)=0$. That's why in Gould and Mays (1984) the auxiliary function is used:

$$g_{s,r}(x)=\left(\frac{1-(1-x)^s}{s\cdot x}\right)^{\frac{1}{s-r}}.$$

Theorem 44. *For the Stolarsky means $\mathcal{D}_{r,s}$, we have the following first terms of the power series expansion*

$$\mathcal{D}_{r,s}(1,1-x)=1-\frac{1}{2}\cdot x+\frac{r+s-3}{24}\cdot x^2+\frac{r+s-3}{48}\cdot x^3$$
$$-\left[2(r^3+r^2s+rs^2+s^3)-5(r+s)^2-70(r+s)+225\right]\cdot\frac{x^4}{5760}$$
$$-\left[2(r^3+r^2s+rs^2+s^3)-5(r+s)^2-30(r+s)+105\right]\cdot\frac{x^5}{3840}$$
$$+[16(r^5+r^4s+r^3s^2+r^2s^3+rs^4+s^5)-42(r^4+2r^3s+2r^2s^2$$
$$+2rs^3+s^4)-1687(r^3+s^3)-1617(r+s)rs+4305(r+s)^2$$
$$+15519(r+s)-59535]\cdot\frac{x^6}{2903040}+\cdots.$$

The series expansion of extended logarithmic means was obtained in Toader et al. (2012).

Theorem 45. *The first coefficients of the series expansion of the extended logarithmic mean $\mathcal{L}_p$ (with $p \neq 0$) are*

$$\mathcal{L}_p(1,1-x) = 1 - \frac{1}{2}x - \frac{p-3}{24}x^2 - \frac{p-3}{48}x^3$$
$$-\frac{(p-5)\left(2p^2+5p-45\right)}{5760}x^4 - \frac{2p^3-5p^2-30p+105}{3840}x^5 + \dots$$

Proof. Denote

$$\mathcal{L}_p(1,1-x) = \sum_{n=0}^{\infty} h_n x^n = \left(\frac{1-(1-x)^p}{-p\ln(1-x)}\right)^{\frac{1}{p}} = \left(\sum_{n=0}^{\infty} l_n x^n\right)^{\frac{1}{p}}. \tag{2.34}$$

This gives

$$1 - \sum_{n=0}^{\infty}(-1)^n \binom{p}{n} x^n = p\left(\sum_{n=1}^{\infty}\frac{x^n}{n}\right)\left(\sum_{n=0}^{\infty} l_n x^n\right).$$

Simplifying with p and x, we deduce the equality

$$\frac{(-1)^n}{n+1}\binom{p-1}{n} = \sum_{k=0}^{n}\frac{l_{n-k}}{k+1},$$

which allows the step-by-step determination of the coefficients l_n,

$$l_0 = 1, \quad l_1 = -\frac{p}{2}, l_2 = \frac{p}{12}(2p-3), \quad l_3 = -\frac{p}{24}(p-2)^2,$$
$$l_4 = \frac{p}{720}\left(6p^3 - 45p^2 + 110p - 90\right),$$
$$l_5 = -\frac{p}{1440}(p-4)^2\left(2p^2 - 8p + 9\right), \dots$$

Using Euler's formula for l_n and h_n,

$$h_n = \frac{1}{n}\sum_{k=1}^{n}\left[k\left(1+\frac{1}{p}\right) - n\right] l_k h_{n-k},$$

we get the result. □

Remark 49. For $q = 0$, the series expansion of the Stolarsky mean $\mathcal{D}_{p,q}$ gives the series expansion of the extended logarithmic mean $\mathcal{L}_p$. For $p = 1$, this gives the series expansion of the logarithmic mean (given in Costin and Toader, 2009), while for $p = 0$, we obtain the series expansion of the geometric mean.

Using series expansion of means, in Lehmer (1971) it was proved that the families of means $\mathcal{P}_q$ and $\mathcal{C}_p$ have in common only the arithmetic, geometric, and harmonic means. More generally, in Costin and Toader (2003) the following result is proved.

Theorem 46. *The families of weighted means $\mathcal{P}_{q;t}$ and $\mathcal{C}_{p;s}$ have in common only the weighted arithmetic mean $\mathcal{A}_t$, the geometric mean $\mathcal{G}$, the weighted harmonic mean $\mathcal{H}_t$, the first and the second projection Π_1 and Π_2.*

In Gould and Mays (1984) the special case of the families of means $\mathcal{D}_{r,s}$ and $\mathcal{C}_p$, which have in common only the arithmetic, geometric, and harmonic means, was considered. More generally, in Costin and Toader (2003) the following result is proved.

Theorem 47. *The families of means $\mathcal{D}_{r,s}$ and $\mathcal{S}_{p,q}$ have in common only the power means.*

2.2.7 Generalized inverses of means

The basic results related to **generalized inverses** of means, that is to complementary with respect to $\mathcal{G}_\lambda$, were given in Costin (2003). We denote the $\mathcal{G}_\lambda$-complementary of M by ${}^{\mathcal{G}(\lambda)}M$. For $\lambda = 1/2$ we use the simpler denotation ${}^{\mathcal{G}}M$.

Theorem 48. *If the mean M has the series expansion*

$$M(1, 1-x) = 1 + \sum_{n=1}^{\infty} a_n x^n,$$

then its generalized inverse ${}^{\mathcal{G}(\lambda)}M$ has the series expansion

$${}^{\mathcal{G}(\lambda)}M(1, 1-x) = 1 + \sum_{n=1}^{\infty} c_n x^n,$$

where the coefficients are given by

$$c_1 = -1 - b_1,$$
$$c_n = -\sum_{k=1}^{n} b_k \cdot c_{n-k}\ ,\ n \geq 2\ ,$$

with

$$b_n = \frac{1}{n}\sum_{k=1}^{n}\left(\frac{k}{1-\lambda} - n\right) \cdot a_k \cdot b_{n-k}\ ,\ n \geq 1, b_0 = 1.$$

Proof. As

$$^{\mathcal{G}(\lambda)}M = \left(\frac{\mathcal{G}_\lambda}{M^\lambda}\right)^{\frac{1}{1-\lambda}},$$

we get

$$^{\mathcal{G}(\lambda)}M\,(1, 1-x) \cdot M^{\frac{\lambda}{1-\lambda}}\,(1, 1-x) = 1-x.$$

Denoting

$$M^{\frac{\lambda}{1-\lambda}}(1, 1-x) = 1 + \sum_{n=1}^{\infty} b_n x^n$$

we get the conditions

$$\begin{cases} c_1 + b_1 = -1 \\ \sum_{k=0}^{n} b_k \cdot c_{n-k} = 0, \; n \geq 2 \end{cases},$$

where b_n is given by Euler's formula. □

Corollary 23. *If the mean M has the series expansion*

$$M(1, 1-x) = 1 + \sum_{n=1}^{\infty} a_n x^n,$$

then the first terms of the series expansion of its generalized inverse $^{\mathcal{G}(\lambda)}M$ are

$$\begin{aligned}
^{\mathcal{G}(\lambda)}M(1, 1-x) &= 1 - (1 + \alpha \cdot a_1) \cdot x + \frac{\alpha}{2}\Big[(\alpha+1) \cdot a_1^2 + 2(a_1 - a_2)\Big] \cdot x^2 \\
&- \frac{\alpha}{6}\Big[(\alpha+1)(\alpha+2) \cdot a_1^3 + 3(\alpha+1) \cdot a_1(a_1 - 2a_2) + 6(a_3 - a_2)\Big] \cdot x^3 \\
&+ \frac{\alpha}{24}\Big[(\alpha+1)(\alpha+2)(\alpha+3) \cdot a_1^4 + 4a_1^2(\alpha+1)(\alpha+2)(a_1 - 3a_2) \\
&+ 12(\alpha+1)\Big(a_2^2 - 2a_1(a_2 - a_3)\Big) + 24(a_3 - a_4)\Big] \cdot x^4 - \frac{\alpha}{5!}[(\alpha+1)(\alpha+2) \\
&\cdot (\alpha+3)(\alpha+4) \cdot a_1^5 + 5a_1^3(\alpha+1)(\alpha+2)(\alpha+3)(a_1 - 4a_2) - 60a_1^2 \\
&\cdot (\alpha+1)(\alpha+2)(a_2 - a_3) + 60a_1(\alpha+1)\Big((\alpha+2)a_2^2 + 2(a_3 - a_4)\Big) \\
&+ 60a_2(\alpha+1)(a_2 - 2a_3) - 120(a_4 - a_5)] \cdot x^5 \\
&+ \frac{\alpha}{6!}\Big[(\alpha+1)(\alpha+2)(\alpha+3)(\alpha+4)(\alpha+5) \cdot a_1^6 + 6a_1^4(\alpha+1)(\alpha+2) \\
&\cdot (\alpha+3)(\alpha+4)(a_1 - 5a_2) - 120a_1^3(\alpha+1)(\alpha+2)(\alpha+3)(a_2 - a_3)
\end{aligned}$$

$$+180(\alpha+1)(\alpha+2)\left((\alpha+3)a_2^2+2(a_3-a_4)\right)+360a_1(\alpha+1)$$
$$\cdot((\alpha+2)(a_2-2a_3)a_2-2(a_4-a_5))-120(\alpha+1)(\alpha+2)a_2^3$$
$$+360(\alpha+1)\cdot\left(a_3^2-2a_2(a_3-a_4)\right)+720(a_5-a_6)\Big]\cdot x^6+\cdots,$$

where

$$\alpha=\frac{\lambda}{1-\lambda}.$$

For the Greek means, the following consequences were given in Toader and Toader (2003a).

Corollary 24. *The first terms of the series expansion of the generalized inverse of the Greek means are:*

$$\mathcal{G}^{(\lambda)}\mathcal{A}(1,1-x)=1+\frac{\alpha-2}{2}\cdot x+\frac{\alpha(\alpha-3)}{8}\cdot x^2$$
$$+\frac{\alpha(\alpha+1)(\alpha-4)}{48}\cdot x^3+\frac{\alpha(\alpha+1)(\alpha+2)(\alpha-5)}{384}\cdot x^4+\cdots;$$
$$\mathcal{G}^{(\lambda)}\mathcal{G}(1,1-x)=1+\frac{\alpha-2}{2}\cdot x+\frac{\alpha(\alpha-2)}{8}\cdot x^2$$
$$+\frac{\alpha\left(\alpha^2-4\right)}{48}\cdot x^3+\frac{\alpha\left(\alpha^2-4\right)(\alpha+4)}{384}\cdot x^4+\cdots;$$
$$\mathcal{G}^{(\lambda)}\mathcal{H}(1,1-x)=1+\frac{\alpha-2}{2}\cdot x+\frac{\alpha(\alpha-1)}{8}\cdot x^2$$
$$+\frac{\alpha(\alpha-1)(\alpha+4)}{48}\cdot x^3+\frac{\alpha(\alpha-1)\left(\alpha^2+11\alpha+22\right)}{384}\cdot x^4+\cdots;$$
$$\mathcal{G}^{(\lambda)}\mathcal{C}(1,1-x)=1+\frac{\alpha-2}{2}\cdot x+\frac{\alpha(\alpha-5)}{8}\cdot x^2$$
$$+\frac{\alpha\left(\alpha^2-9\alpha-4\right)}{48}\cdot x^3+\frac{\alpha(\alpha-1)\left(\alpha^2-13\alpha-26\right)}{384}\cdot x^4+\cdots;$$
$$\mathcal{G}^{(\lambda)}\mathcal{F}_5(1,1-x)=1+\frac{\alpha-2}{2}\cdot x+\frac{\alpha(\alpha-4)}{8}\cdot x^2$$
$$+\frac{\alpha(\alpha+1)(\alpha-7)}{48}\cdot x^3+\frac{\alpha\left(\alpha^3-8\alpha^2-28\alpha-256\right)}{384}\cdot x^4+\cdots;$$
$$\mathcal{G}^{(\lambda)}\mathcal{F}_6(1,1-x)=1+\frac{\alpha-2}{2}\cdot x+\frac{\alpha(\alpha-4)}{8}\cdot x^2$$
$$+\frac{\alpha\left(\alpha^2-6\alpha-1\right)}{48}\cdot x^3+\frac{\alpha\left(\alpha^3-8\alpha^2-4\alpha+8\right)}{384}\cdot x^4+\cdots;$$
$$\mathcal{G}^{(\lambda)}\mathcal{F}_7(1,1-x)=1+(\alpha-1)\cdot x+\frac{\alpha(\alpha-3)}{2}\cdot x^2$$

$$+\frac{\alpha\left(\alpha^2-6\alpha-1\right)}{6}\cdot x^3+\frac{2\alpha\left(\alpha+1\right)\left(\alpha^2-3\alpha-1\right)}{3}\cdot x^4+\cdots;$$

$$^{\mathcal{G}(\lambda)}\mathcal{F}_8(1,1-x)=1+(\alpha-1)\cdot x+\frac{\alpha\left(\alpha-3\right)}{2}\cdot x^2$$

$$+\frac{\alpha\left(\alpha-1\right)\left(\alpha-5\right)}{6}\cdot x^3+\frac{\alpha\left(\alpha-1\right)\left(\alpha-2\right)\left(\alpha-7\right)}{24}\cdot x^4+\cdots;$$

$$^{\mathcal{G}(\lambda)}\mathcal{F}_9(1,1-x)=1-x+\alpha\cdot x^2-\alpha\cdot x^3+\frac{\alpha\left(\alpha+1\right)}{2}\cdot x^4+\cdots,$$

respectively

$$^{\mathcal{G}(\lambda)}\mathcal{F}_{10}(1,1-x)=1-x+\alpha\cdot x^2-2\alpha\cdot x^3+\frac{\alpha\left(2\alpha+11\right)}{4}\cdot x^4+\cdots$$

Using them we can deduce the following property:

Corollary 25. *We have*

$$^{\mathcal{G}(\lambda)}\mathcal{F}_i=\mathcal{F}_j,\ \textit{with } i,j\in\{1,2,...,10\},$$

in and only in the cases

$$^{\mathcal{G}}\mathcal{A}=\mathcal{H},\ ^{\mathcal{G}}\mathcal{F}_8=\mathcal{F}_9,$$

or their duals.

The series expansion of the generalized inverse of $\mathcal{S}_{p,p-r;\mu}$ was given in Costin (2004).

Corollary 26. *The first terms of the series expansion of the generalized inverse of $\mathcal{S}_{p,p-q;\mu}$ are*

$$\begin{aligned}
^{\mathcal{G}(\lambda)}\mathcal{S}_{p,p-q;\mu}(1,1-x)&=1-(\alpha\mu-\alpha+1)\cdot x-\alpha(1-\mu)[(\alpha+2p-q)\mu\\
&-(\alpha-1)]\cdot\frac{x^2}{2!}+\alpha(1-\mu)\{[6p^2+6(\alpha-q)p+(\alpha-q)(\alpha-2q)]\mu^2\\
&-[3p^2-3(q-2\alpha)p+(2\alpha-q)(\alpha-q)]\mu+(\alpha-1)(\alpha+1)\}\cdot\frac{x^3}{3!}\\
&-\alpha(1-\mu)\{[24p^3+36(\alpha-q)p^2+12(\alpha-q)(\alpha-2q)p\\
&+(\alpha-q)(\alpha-2q)(\alpha-3q)]\mu^3+[-24p^3+12(3q-4\alpha-1)\\
&\cdot p^2-12(2\alpha-2q+1)(\alpha-q)p-(\alpha-2q)(\alpha-q)(3\alpha+2-3q)]\mu^2+[4p^3\\
&+6(2\alpha-q+1)p^2+2(6\alpha(2\alpha-2q+1)-3q+2q^2-1)p\\
&+(\alpha-q)(3\alpha^2+4\alpha-3q\alpha-2q+q^2-1)]\mu-(\alpha-1)(\alpha+1)(\alpha+2)\}\cdot\frac{x^4}{4!}
\end{aligned}$$

$$+\alpha(1-\mu)\{[120p^4+240(\alpha-q)p^3+120(\alpha-q)(\alpha-2q)p^2$$
$$+20(\alpha-q)(\alpha-2q)(\alpha-3q)p+(\alpha-q)(\alpha-2q)(\alpha-3q)(\alpha-4q)]\mu^4$$
$$+[-180p^4+60(6q-7\alpha-2)p^3-90(\alpha-q)(3\alpha-4q+2)p^2$$
$$-30(\alpha-q)(\alpha-2q)(2\alpha+2-3q)p-(\alpha-q)(\alpha-2q)(\alpha-3q)(4\alpha+5$$
$$-6q)]\mu^3+[70p^4+20(10\alpha-7q+6)p^3+10(-30q\alpha+18\alpha^2+24\alpha+3$$
$$+14q^2-18q)p^2+10(\alpha-q)(6\alpha^2+12\alpha-12q\alpha+7q^2-12q+3)p$$
$$+(6\alpha^2-12q\alpha+15\alpha+5+7q^2-15q)(\alpha-2q)(\alpha-q)]\mu^2+[-5p^4$$
$$+10(q-2-2\alpha)p^3+(30q\alpha-30\alpha^2-60\alpha-15-10q^2+30q)p^2$$
$$-52\alpha+(2-q)(2\alpha^2-2q\alpha+4\alpha-2q+q^2-1)p-(\alpha-q)(4\alpha^3-6q\alpha^2$$
$$+15\alpha^2-15q\alpha+10\alpha+4q^2\alpha-5+5q^2-q^3-5q)]\mu$$
$$+(\alpha-1)(\alpha+1)(\alpha+2)(\alpha+3)\}\cdot\frac{x^5}{5!}+\cdots,$$

where $\alpha=\frac{\lambda}{1-\lambda}$.

Remark 50. For $q=p$ we get the first terms of the series of ${}^{\mathcal{G}(\lambda)}\mathcal{P}_{p;\mu}(1,1-x)$, while for $q=1$ we have also the first terms of the series of ${}^{\mathcal{G}(\lambda)}\mathcal{C}_{p;\mu}(1,1-x)$.

Using the above results, the following property was proved in Costin and Toader (2006b).

Theorem 49. *The relation*

$${}^{\mathcal{G}(\lambda)}\mathcal{S}_{p,p-q;\mu}=\mathcal{S}_{r,r-s;\nu}$$

holds if and only if we are in one of the following cases:

(i) ${}^{\mathcal{G}(0)}\mathcal{S}_{p,p-q;\mu}=\mathcal{S}_{r,r-s;0}$;
(ii) ${}^{\mathcal{G}(\lambda)}\mathcal{S}_{p,p-q;1}=\mathcal{S}_{r,r-s;0}$;
(iii) ${}^{\mathcal{G}}\mathcal{S}_{p,p-q;0}=\mathcal{S}_{r,r-s;1}$;
(iv) ${}^{\mathcal{G}(1/3)}\mathcal{S}_{p,p-q;0}=\mathcal{S}_{r,-r}$;
(v) ${}^{\mathcal{G}}\mathcal{S}_{p,-p}=\mathcal{S}_{r,-r}$;
(vi) ${}^{\mathcal{G}(2/3)}\mathcal{S}_{p,-p}=\mathcal{S}_{r,r-s;1}$;
(vii) ${}^{\mathcal{G}}\mathcal{S}_{p,p-q;\mu}=\mathcal{S}_{-p,-p+q;1-\mu}$,

or in an equivalent case, taking into account the property $\mathcal{S}_{s,r;\nu}=\mathcal{S}_{r,s;\nu}$.

Proof. Equating the coefficients of x, in ${}^{\mathcal{G}(\lambda)}\mathcal{S}_{p,p-q;\mu}(1,1-x)$ and in $\mathcal{S}_{r,r-s;\nu}(1,1-x)$ we have the condition

$$\nu=\alpha(1-\mu). \tag{2.35}$$

Then, the equality of the coefficients of x^2 gives the condition

$$\alpha(1-\mu)\left[\mu(2p-q)+(1-\alpha+\alpha\mu)(2r-s)\right]=0. \tag{2.36}$$

We consider the following cases: 1) $\alpha=0$, which gives $\nu=0$ and so the equality (i); 2) $\mu=1$, which also gives $\nu=0$ and so the equality (ii); 3) For $\mu=0$, (2.35) and (2.36) imply $\nu=\alpha$ and $\alpha(1-\alpha)(2r-s)=0$. So we have to consider the following special cases: 3.1) $\mu=\nu=\alpha=0$ which represents a special case of (i); 3.2) $\mu=0$, $\nu=\alpha=1$ leading to (iii); 3.3) $\mu=0, s=2r$, which, assuming also the equality of the coefficients of x^3, gives $\alpha=\nu=1/2$, a special case of (iv). 4) For $\mu\neq 0$, (2.36) implies

$$p=\frac{(1-\alpha+\alpha\mu)(r-2q)+\mu}{2\mu}. \tag{2.37}$$

Replacing (2.35) and (2.37) in the coefficients of x^3, we get again some special cases: 4.1) $\mu=\nu=1/2, \alpha=1, 2p=q+r+s-2r$; 4.2) $\alpha=2, \nu=1, \mu=1/2, q=2p$ which leads to (vi); or 4.3)

$$\begin{aligned}&\left[2\alpha^2\mu^3+\alpha\mu^2(7-8\alpha)+5\mu\left(2\alpha^2-3\alpha+1\right)-4(\alpha-1)^2\right]r^2\\&+\left[2\alpha\mu^3(2p-s\alpha)+\mu^2\left(8s\alpha^2-6p\alpha-7s\alpha+4p\right)+\mu(15s\alpha-10s\alpha^2\right.\\&\left.+2p\alpha-2p-5\alpha)+4s(\alpha-1)^2\right]r+2p\mu^3(p-s\alpha)+\mu^2\left(s^2\alpha-s^2\alpha^2-2ps\right.\\&\left.-p^2+3ps\alpha\right)+\mu\left(2s^2\alpha^2-3s^2\alpha-ps\alpha+ps+s^2\right)-s^2(\alpha-1)^2=0.\end{aligned}$$

In the cases 4.1) we pass to the coefficients of x^4 which gives

$$(q-2p)(p+r)(p+r-q)=0,$$

getting so a new split: 4.1.1) $p=-r, q=-s$ giving a special case of (vii); 4.1.2) $q=2p, s=2r$, giving (vi); 4.1.3) $q=s, p=s-r$, leading to a special case of (vii). Similarly, in the case 4.3), the equality of the coefficients of x^4 gives the split: 4.3.1) $\alpha=1$, $p=-r, q=-s$, $\nu=1-\mu$, thus (vii); 4.3.2) $\alpha=1$, $p=s-r, q=s, \nu=1-\mu$ giving also (vii); 4.3.3) $s=2r$; 4.3.4)

$$\alpha=\frac{s^2\left(2\mu^2-5\mu+1\right)-2r(s-r)\left(8\mu^2-11\mu+2\right)}{(\mu-1)\left[s^2(4\mu-1)+4r(s-r)\left(3\mu^2-5\mu+1\right)\right]}.$$

To solve the case 4.3.3) we pass to the coefficients of x^5. We fall on special cases giving again (iv)–(vii). In the case 4.3.4) the coefficients of x^5 give the condition $s=zr$, where z is a root of the equation: $\mu z^2+(z-1)\left(6\mu^2-8\mu+1\right)=0$. In this case we have to pass also to the coefficients of x^6, but we get no new solution. The validity of the cases (i)–(vii) can be verified directly. □

Corollary 27. *The mean $\mathcal{G}_\lambda$ is invariant with respect to the set of weighted Gini means only in the following non-trivial cases: (2.23), (2.24), and*

$$\mathcal{G}\left(\mathcal{S}_{p,q;\mu}, \mathcal{S}_{-p,-q;1-\mu}\right) = \mathcal{G}. \tag{2.38}$$

Some special results were also proved in Costin (2004), Costin and Toader (2005, 2005a). Similar results were given in Toader (2007).

Theorem 50. *The relation*

$$^{\mathcal{G}(\lambda)}\mathcal{S}_{p,p-q;\mu} = \mathcal{D}_{r,s}$$

holds if and only if we are in one of the following non-trivial cases:

$$(i)\ ^{\mathcal{G}(1/3)}\mathcal{S}_{p,p-q;0} = \mathcal{D}_{r,-r};$$
$$(ii)\ ^{\mathcal{G}}\mathcal{S}_{p,0} = \mathcal{D}_{-p,-2p}.$$

Theorem 51. *The relation*

$$^{\mathcal{G}(\lambda)}\mathcal{D}_{r,s} = \mathcal{S}_{p,q;\mu}$$

holds if and only if we are in one of the following non-trivial cases:

$$(i)\ ^{\mathcal{G}(2/3)}\mathcal{D}_{r,-r} = \mathcal{S}_{p,q;1};$$
$$(ii)\ ^{\mathcal{G}}\mathcal{D}_{2s,s} = \mathcal{S}_{0,-s}.$$

Theorem 52. *The relation*

$$^{\mathcal{G}(\lambda)}\mathcal{D}_{r,s} = \mathcal{D}_{p,q}, rspq(r-s)(p-q) \neq 0$$

holds if and only if we are in the following non-trivial case:

$$^{\mathcal{G}}\mathcal{D}_{r,s} = \mathcal{D}_{-r,-s}.$$

Remark 51. For $\lambda = 1/2$, the complete result, thus including $rspq(r-s)(p-q) = 0$, was proved otherwise in Błasińska-Lesk et al. (2003).

Theorem 53. *The relation*

$$^{\mathcal{G}}\mathcal{D}_{r,s} = \mathcal{D}_{p,q}$$

holds if and only if we are in one of the following cases:

$$(i)\ ^{\mathcal{G}}\mathcal{D}_{0,0} = \mathcal{D}_{0,0};$$
$$(ii)\ ^{\mathcal{G}}\mathcal{D}_{r,0} = \mathcal{D}_{-r,0};$$
$$(iii)\ ^{\mathcal{G}}\mathcal{D}_{r,r} = \mathcal{D}_{-r,-r};$$
$$(iv)\ ^{\mathcal{G}}\mathcal{D}_{r,-r} = \mathcal{D}_{p,-p},\ or\ ^{\mathcal{G}}\mathcal{D}_{r,s} = \mathcal{D}_{-r,-s},\ or\ ^{\mathcal{G}}\mathcal{D}_{r,s} = \mathcal{D}_{-s,-r}.$$

2.2.8 Complementariness with respect to power means

Basic results related to complementariness with respect to power means were given in Costin (2004b). Denote the $\mathcal{P}_{m;\lambda}$-complementary of M by ${}^{\mathcal{P}(m;\lambda)}M$, or by ${}^{\mathcal{P}(m)}M$ if $\lambda = 1/2$.

Theorem 54. *If the mean M has the series expansion*

$$M(1, 1-x) = 1 + \sum_{n=1}^{\infty} a_n x^n,$$

then ${}^{\mathcal{P}(m;\lambda)}M$ *has, for* $m \neq 0$ *and* $\lambda \neq 0, 1$, *the series expansion*

$${}^{\mathcal{P}(m;\lambda)}M(1, 1-x) = 1 + \sum_{n=1}^{\infty} c_n x^n,$$

where

$$c_n = \frac{1}{n} \sum_{k=1}^{n} \left[k\left(\frac{1}{m}+1\right) - n \right] \cdot d_k \cdot c_{n-k} \, , \; n \geq 1,$$

with

$$d_n = (-1)^n \frac{(m)_n}{n!} - \alpha b_n \, , \; n \geq 1 \, ,$$

$$b_n = \frac{1}{n} \sum_{k=1}^{n} [k(m+1) - n] \cdot a_k \cdot b_{n-k} \, , \; n \geq 1,$$

and

$$b_0 = 1, \; \alpha = \frac{\lambda}{1-\lambda},$$

where $(m)_n$ *is given by (1.30).*

Proof. As

$${}^{\mathcal{P}(m;\lambda)}M = \left(\frac{\mathcal{P}_{m;\lambda} - \lambda M^m}{1-\lambda} \right)^{\frac{1}{m}},$$

we get

$$(1-\lambda) \cdot \left[{}^{\mathcal{P}(m;\lambda)}M \right]^m (1, 1-x) + \lambda \cdot M^m (1, 1-x) = \lambda + (1-\lambda)(1-x)^m .$$

Denoting

$$M^m(1,1-x)=1+\sum_{n=1}^{\infty} b_n x^n$$

and

$$\left[{}^{\mathcal{P}(m;\lambda)}M\right]^m(1,1-x)=1+\sum_{n=1}^{\infty} d_n x^n,$$

we have

$$(1-\lambda)\cdot\sum_{n=1}^{\infty} d_n x^n+\lambda\cdot\sum_{n=1}^{\infty} b_n x^n=(1-\lambda)\cdot\sum_{n=1}^{\infty}(-1)^n\frac{(m)_n}{n!}x^n.$$

Putting $\alpha=\lambda/(1-\lambda)$, we get

$$d_n=(-1)^n\frac{(m)_n}{n!}-\alpha b_n\ ,\ n\geq 1,$$

where

$$b_n=\frac{1}{n}\sum_{k=1}^{n}[k(m+1)-n]\cdot a_k\cdot b_{n-k}\ ,\ n\geq 1.$$

Taking

$${}^{\mathcal{P}(m;\lambda)}M(1,1-x)=1+\sum_{n=1}^{\infty} c_n x^n=\left(1+\sum_{n=1}^{\infty} d_n x^n\right)^{\frac{1}{m}},$$

we have

$$c_n=\frac{1}{n}\sum_{k=1}^{n}\left[k\left(\frac{1}{m}+1\right)-n\right]\cdot d_k\cdot c_{n-k}\ ,\ n\geq 1,$$

obtaining the desired formulas. □

Corollary 28. *If the mean M has the series expansion*

$$M(1,1-x)=1+\sum_{n=0}^{\infty} a_n x^n,$$

then the first terms of the series expansion of ${}^{\mathcal{P}(m;\lambda)}M$, for $m\neq 0$ and $\lambda\neq 0,1$, are

$${}^{\mathcal{P}(m;\lambda)}M(1,1-x)=1-(1+\alpha\cdot a_1)\cdot x+\frac{\alpha}{2}[(1-m)(2a_1+a_1^2+\alpha a_1^2)$$

$$-2a_2]\cdot x^2+\frac{\alpha}{6}\{(1-m)[(2\alpha^2m-\alpha^2+3\alpha m-3\alpha+m-2)a_1^3$$
$$+3(2\alpha m-\alpha+m-1)a_1^2+3ma_1+6a_2+6(\alpha+1)a_1a_2]-6a_3\}\cdot x^3$$
$$+\frac{\alpha}{24}\{(1-m)\cdot[(6\alpha^3m^2-5\alpha^3m+\alpha^3+12\alpha^2m^2-18\alpha^2m+6\alpha^2+7\alpha m^2$$
$$-18\alpha m+11\alpha+m^2-5m+6)a_1^4+4(6\alpha^2m^2-5\alpha^2m+\alpha^2+6\alpha m^2-9\alpha m$$
$$+3\alpha+m^2-3m+2)a_1^3+6(4\alpha m^2-2\alpha m+m^2-m)a_1^2+4m(m+1)a_1$$
$$+12ma_2+24(2\alpha m-\alpha+m-1)a_1a_2+12(2\alpha^2m-\alpha^2+3\alpha m-3\alpha$$
$$+m-2)a_1^2a_2+24(\alpha+1)a_1a_3+24a_3+12(\alpha+1)a_2^2]-24a_4\}\cdot x^4+...,$$

where

$$\alpha=\frac{\lambda}{1-\lambda}.$$

Using them, the following consequence was proved in Toader and Toader (2004b).

Corollary 29. *We have*

$$^{\mathcal{P}(m;\lambda)}\mathcal{F}_i=\mathcal{F}_j,\ with\ i,j\in\{1,2,...,10\},$$

in and only in the cases

$$^{\mathcal{A}}\mathcal{H}=\mathcal{C},\ ^{\mathcal{A}}\mathcal{F}_7=\mathcal{F}_9,$$

or their duals.

Proof. Lemma 11 implies $\lambda=1/2$, and then equating the coefficients of x^k $(k=1,2,...)$ in $^{\mathcal{P}(m)}\mathcal{F}_i(1,1-x)$ and in $\mathcal{F}_j(1,1-x)$, we get the desired result. □

Another consequence was proved in Costin (2007a).

Theorem 55. *The first terms of the series expansion of the $\mathcal{P}_{m;\lambda}$-complementary of $\mathcal{P}_{p;\mu}$ are*

$$^{\mathcal{P}(m;\lambda)}\mathcal{P}_{p;\mu}(1,1-x)=1-(\alpha\mu-\alpha+1)x-\frac{\alpha}{2}(\mu-1)(\mu m-\alpha\mu+\alpha\mu m$$
$$-\mu p+\alpha-1-\alpha m+m)x^2-\frac{\alpha}{6}(\mu-1)(3\alpha\mu^2m^2-3\mu^2mp-3\alpha\mu^2m$$
$$-3\alpha^2\mu^2m-3\alpha\mu^2mp+3\alpha\mu^2p+2\alpha^2\mu^2m^2+\alpha^2\mu^2+\mu^2m^2+2\mu^2p^2$$
$$+3\alpha\mu mp-\mu p^2-4\alpha^2\mu m^2-3\alpha\mu p+6\alpha^2\mu m-2\alpha^2\mu+\mu m^2-1-3\alpha^2m$$
$$+\alpha^2-3\alpha m^2+3\alpha m+2\alpha^2m^2+m^2)x^3-\frac{\alpha}{24}(\mu-1)(-2-\alpha-\alpha m\mu$$

$$
\begin{aligned}
&+6\alpha^2 m\mu + 6\alpha^2 m\mu^3 - 18\alpha^2 m^2\mu^3 - 6\alpha\mu p - 12\alpha^2 m\mu^2 + 10\alpha^2 m^2\mu \\
&-33\alpha^3 m^2\mu - 6\alpha m\mu^2 + 3\alpha m^2\mu^2 - 6m\mu^2 p + 6\alpha\mu^2 p + 22\alpha^2 m^2\mu^2 \\
&+33\alpha^3 m^2\mu^2 + 3\alpha m^2\mu - 11\alpha^3 m^2\mu^3 - 7\alpha m^2\mu^3 + 18\alpha^3 m\mu - 18\alpha^3 m\mu^2 \\
&+6\alpha^3 m\mu^3 - 6\alpha^2\mu p + 12\alpha^2\mu^2 p - 6\alpha^2\mu^3 p + 2\alpha^2 - 11\alpha\mu^3 p^2 + 15\alpha\mu^2 p^2 \\
&-4\alpha\mu p^2 + 7\alpha m - m + 18m\alpha\mu^3 p - 18m\alpha\mu^2 p + 2\alpha^2\mu^2 - 4\alpha^2\mu + \alpha\mu - m\mu \\
&+\mu p + \alpha m^2 - 14\alpha^2 m^2 + 2m^2 + 2m^2\mu^2 + 2m^2\mu - 2\mu p^2 + 4\mu^2 p^2 \\
&+18\alpha^2 m\mu p + 18\alpha^2 m\mu^3 p - 36\alpha^2 m\mu^2 p + \alpha^3 + 11\alpha^3 m^2 - 6\alpha^3 m - \alpha^3\mu^3 \\
&+3\alpha^3\mu^2 - 3\alpha^3\mu + 6m^2\alpha\mu p - 18m^2\alpha\mu^3 p + 12m^2\alpha\mu^2 p - 12\alpha^2 m^2\mu p \\
&-12\alpha^2 m^2\mu^3 p + 24\alpha^2 m^2\mu^2 p + 11m\alpha\mu^3 p^2 - 15m\alpha\mu^2 p^2 + 4m\alpha\mu p^2 - 7\alpha m^3 \\
&+12\alpha^2 m^3 - 6\alpha^3 m^3 - \mu p^3 + 6\mu^2 p^3 - 6\mu^3 p^3 + 12\alpha^2 m^3\mu^3 - 6m^2\mu^3 p \\
&-12\alpha^2 m^3\mu + 18\alpha^3 m^3\mu + 3\alpha m^3\mu^2 - 18\alpha^3 m^3\mu^2 - 3\alpha m^3\mu + 6\alpha^3 m^3\mu^3 \\
&+7\alpha m^3\mu^3 - 7m\mu^2 p^2 + 11m\mu^3 p^2 + m^3\mu + m^3\mu^2 + m^3\mu^3 + m^3 \\
&-12\mu^2\alpha^2 m^3)x^4 + \cdots,
\end{aligned}
$$

where

$$\alpha = \frac{\lambda}{1-\lambda}.$$

The problem of invariance in the family of weighted power means was solved in Costin (2007).

Theorem 56. *We have*

$$^{\mathcal{P}(m;\lambda)}\mathcal{P}_{p;\mu} = \mathcal{P}_{q;\nu}, m \neq 0,$$

if and only if we are in one of the non-trivial cases:

$$^{\mathcal{P}(m;\lambda)}\mathcal{P}_{p;0} = \mathcal{P}_{m;\lambda/(1-\lambda)}; \tag{2.39}$$

$$^{\mathcal{P}(m;\lambda)}\mathcal{P}_{m;\mu} = \mathcal{P}_{m;\lambda(1-\mu)/(1-\lambda)}; \tag{2.40}$$

$$^{\mathcal{P}(m;\lambda)}\mathcal{P}_{m;(2\lambda-1)/\lambda} = \mathcal{P}_{q;1}; \tag{2.41}$$

$$^{\mathcal{P}(2p;(1+\mu)/2)}\mathcal{P}_{p;\mu} = \mathcal{P}_{p;1+\mu}; \tag{2.42}$$

$$^{\mathcal{P}(2p;1/5)}\mathcal{P}_{p;-1} = \mathcal{P}_0; \tag{2.43}$$

$$^{\mathcal{P}(2q;4/5)}\mathcal{P}_0 = \mathcal{P}_{q;2}. \tag{2.44}$$

Proof. Equating the coefficients of $x, x^2, ..., x^5$ in $^{\mathcal{P}(m;\lambda)}\mathcal{P}_{p;\mu}(1, 1-x)$ and in $\mathcal{P}_{q;\nu}(1, 1-x)$, the above non-trivial solutions were obtained. □

Remark 52. Some of the complementaries in the above theorem are only pre-means.

Corollary 30. *We have*

$$^{\mathcal{P}(m;\lambda)}\mathcal{P}_{p;\mu} = \mathcal{P}_{q;\nu}, m \neq 0,$$

where $\mathcal{P}_{q;\nu}$ is a mean, if and only if we are in one of the non-trivial cases:

$$(i)\ ^{\mathcal{P}(m;\lambda)}\mathcal{P}_{p;0} = \mathcal{P}_{m;\lambda/(1-\lambda)};\ \lambda \in [0, 1/2];$$
$$(ii)\ ^{\mathcal{P}(m;\lambda)}\mathcal{P}_{m;\mu} = \mathcal{P}_{m;\lambda(1-\mu)/(1-\lambda)};\ \lambda \in [0, 1/(2-\mu)];$$
$$(iii)\ ^{\mathcal{P}(m;\lambda)}\mathcal{P}_{m;(2\lambda-1)/\lambda} = \mathcal{P}_{q;1};\ \lambda \in [1/2, 1].$$

Remark 53. The problem of invariance in the class of weighted quasi-arithmetic means was solved by another method in Jarczyk and Matkowski (2006) and Jarczyk (2007). Of course, the weighted power means are weighted quasi-arithmetic means, but the above results include pre-means as complementaries. The problem of invariance in the class of (symmetric) power means was solved in Lehmer (1971). The problem of reproducing identities for power means,

$$\mathcal{P}_m(\mathcal{P}_p, \mathcal{P}_q) = \mathcal{P}_r,$$

was solved in Brenner and Mays (1987). Only the trivial solution,

$$\mathcal{P}_m(\mathcal{P}_p, \mathcal{P}_p) = \mathcal{P}_p,$$

and the solutions of the invariance problem,

$$\mathcal{P}_m(\mathcal{P}_p, \mathcal{P}_q) = \mathcal{P}_m,$$

exist.

Remark 54. Some cases in which $^{\mathcal{P}(m;\lambda)}\mathcal{P}_{p;\mu} = \mathcal{S}_{q,r;\nu}$ are given in Costin (2007a).

The problem of invariance of a weighted power mean with respect to the set of weighted Gini means was studied in Costin and Toader (2012b). First of all, from Corollary 28 it is deduced:

Corollary 31. *The first terms of the series expansion of the $\mathcal{P}_{p;\lambda}$-complementary of $\mathcal{S}_{r,r-s;\mu}$ are*

$$^{\mathcal{P}(p;\lambda)}\mathcal{S}_{r,r-s;\mu}(1, 1-x) = 1 - \frac{1-2\lambda+\lambda\mu}{1-\lambda}x - \frac{\lambda(1-\mu)}{2(1-\lambda)^2}(\lambda\mu - 2r\mu\lambda + sp\lambda$$

$$-p\mu - s\mu + 2r\mu - 2p\lambda - p - 2\lambda - 1)x^2 + \frac{\lambda(1-\mu)}{6(1-\lambda)^3}(p^2 - 3p^2\mu\lambda^2$$
$$+p^2\mu^2\lambda + 3rs\mu - s^2\mu - 3r^2\mu + 6r^2\mu^2 - 6rs\mu^2 + 2s^2\mu^2 + 2\lambda + 6pr\lambda\mu$$
$$-3ps\lambda\mu + 6pr\lambda\mu^2 - 3ps\lambda\mu^2 - 6p\lambda^2 - 5p^2\lambda + 6p^2\lambda^2 - 6pr\lambda^2\mu + 3ps\lambda^2\mu$$
$$+\mu^2\lambda^2 + 3p\lambda + 2s^2\mu\lambda + 3rs\mu\lambda^2 - 6sr\lambda^2\mu^2 - 6rs\mu\lambda + 12rs\lambda\mu^2 - 6r^2\lambda\mu$$
$$+p^2\mu^2 + p^2\mu - 12r^2\mu^2\lambda - 4s^2\mu^2\lambda - s^2\mu\lambda^2 - 3r^2\mu\lambda^2 + 6r^2\mu^2\lambda^2$$
$$+2s^2\mu^2\lambda^2 - 6pr\mu^2 - 3ps\mu^2 - 2p^2\mu\lambda - 2\lambda^2\mu - 3p\mu^2\lambda + 6p\mu\lambda^2 - 6p\mu\lambda$$
$$+3s\mu\lambda + 6r\mu^2\lambda - 3s\mu^2\lambda + 6r\mu\lambda^2 - 3s\mu\lambda^2 - 6r^2\mu\lambda^2 + 3s\mu^2\lambda^2 - 1)x^3 + \ldots$$

Using it, the main result was proved:

Theorem 57. *We have:*

$$^{\mathcal{P}(p;\lambda)}\mathcal{S}_{r,m;\mu} = \mathcal{S}_{u,t;v}$$

if we are in one of the non-trivial cases (2.20), (2.21), (2.22), (2.23), (2.38), (2.39), (2.40), (2.41), (2.42), (2.43), (2.44) or

$$(i)\ ^{\mathcal{P}(p;\lambda)}\mathcal{P}_{p;(3\lambda-1)/2} = \mathcal{P}_p$$

and

$$(ii)\ ^{\mathcal{P}(p)}\mathcal{S}_{r,r+p;v} = \mathcal{S}_{-r,p-r;1-v},$$

as well as in its special cases

$$(iii)\ ^{\mathcal{P}(p)}\mathcal{S}_{r-p,r;\mu} = \mathcal{S}_{p-r,2p-r;1-\mu}$$
$$(iv)\ ^{\mathcal{P}(p)}\mathcal{S}_{r,r+p} = \mathcal{S}_{-r,p-r}$$
$$(v)\ ^{\mathcal{P}(p)}\mathcal{S}_{r-p;r} = \mathcal{S}_{p-r,2p-r}$$
$$(vi)\ ^{\mathcal{P}(2r)}\mathcal{S}_{r,3r} = \mathcal{G}$$
$$(vii)\ ^{\mathcal{P}(2r)}\mathcal{G} = \mathcal{S}_{r,3r}.$$

Remark 55. The proof is similar to that of Theorem 56 but the coefficients are much more complicated. For this it was necessary to use a computer algebra system. However we are not sure that "if" in the enunciation of the previous theorem can be replaced by "if and only if," taking into account the warning issued by the computer algebra system, that solutions may have been lost in solving some systems of equations.

2.2.9 Complementariness with respect to Lehmer means

Denote the $C_{p;\lambda}$-complementary of the mean M by ${}^{C(p;\lambda)}M$, or by ${}^{C(p)}M$ if $\lambda = 1/2$. Using Euler's formula, the following result was established in Toader and Toader (2006).

Theorem 58. *If the mean M has the series expansion*

$$M(1, 1-x) = 1 + \sum_{n=0}^{\infty} a_n x^n,$$

then the first terms of the series expansion of ${}^{C(p;\lambda)}M$, *for* $\lambda \neq 0, 1$, *are*

$$\begin{aligned}
{}^{C(p;\lambda)}M(1,1-x) &= 1 - \frac{1-\lambda+\lambda a_1}{1-\lambda}x - \frac{\lambda}{(1-\lambda)^2}[(p-1)a_1(a_1+2(1-\lambda)) \\
&+a_2(1-\lambda)]\cdot x^2 - \frac{\lambda}{2(1-\lambda)^3}[a_1(p-1)(2\lambda^3 p - \lambda^2(p+2) - 4\lambda(p-1) \\
&+3p-2) + a_1^2(p-1)(2\lambda^2(1-3p) + \lambda(3p+2) + 3p - 4) \\
&+a_1^3(p-1)(2\lambda p + p - 2) + 4a_2(p-1)(1-\lambda)^2 + 4a_1 a_2(p-1)(1-\lambda) \\
&+2a_3(1-\lambda)^2]\cdot x^3 + \cdots \\
M^{C(q;\nu)}(1,1-x) &= 1 - (1+\alpha\cdot a_1)\cdot x - \alpha[a_1^2(\alpha q - \alpha + q - 1) - 2a_1 \\
&+a_2]\cdot x^2 - \frac{\alpha}{2(1+\alpha)}[a_1(2\alpha - 5q - 7\alpha q + 2 + 3q^2 + 5\alpha q^2) \\
&+a_1^2(10\alpha - 15q\alpha^2 - 10\alpha q + 6\alpha^2 + 4 - 7q - 12q\alpha + 9q^2\alpha^2 + 12q^2\alpha + 3q^2) \\
&+a_1^3(2 + 6\alpha + 6\alpha^2 + \alpha^3 - 3q - 11q\alpha - 13q\alpha^2 - 5q\alpha^3 + 5q^2\alpha + 7q^2\alpha^2 \\
&+3q^2\alpha^3 + q^2) + 2a_2(1+\alpha)(2q - r - 1) + 2a_1 a_2(1+\alpha)^2(2q - r - 1) \\
&+2a_3(1+\alpha)]\cdot x^3 + \cdots,
\end{aligned}$$

where $\alpha = \nu/(1-\nu)$.

Corollary 32. *We have*

$$\mathcal{C}_{q;\nu}(\mathcal{F}_i, \mathcal{F}_j) = \mathcal{C}_{q;\nu}$$

if and only if $\nu = 1/2$ *and we are in one of the following non-trivial cases:*

$$(i)\ \mathcal{C}_{1/2}(\mathcal{A}, \mathcal{H}) = \mathcal{C}_{1/2};$$
$$(ii)\ \mathcal{C}_1(\mathcal{H}, \mathcal{C}) = \mathcal{C}_1;$$
$$(iii)\ \mathcal{C}_1(\mathcal{F}_7, \mathcal{F}_9) = \mathcal{C}_1;$$
$$(iv)\ \mathcal{C}_{1/2}(\mathcal{F}_8, \mathcal{F}_9) = \mathcal{C}_{1/2}.$$

The invariance in the family of weighted Lehmer means was studied in Costin and Toader (2008).

Corollary 33. *The first terms of the series expansion of* ${}^{C(p;\lambda)}C_{r;\mu}$ *are*

$$
\begin{aligned}
{}^{C(p;\lambda)}C_{r;\mu}(1,1-x) &= 1 - \frac{1-2\lambda+\lambda\mu}{1-\mu}x + \frac{\lambda(1-\mu)}{(1-\lambda)^2}[p(1-2\lambda+\mu) \\
&\quad +\mu r(\lambda-1) - 1 + 2\lambda - \lambda\mu]x^2 + \\
&\frac{\lambda(1-\mu)}{(1-\lambda)^3}\Big[p^2\left(2\lambda^3+2\lambda\mu^2-6\lambda^2\mu-\lambda\mu+5\lambda^2+\mu^2+\mu-5\lambda+1\right) \\
&+4pr\left(\lambda\mu^2+\lambda\mu-\lambda^2\mu-\mu^2\right)+r^2\left(2\lambda\mu-4\lambda\mu^2-\lambda^2\mu-\mu+2\mu^2\right) \\
&\quad +p\Big(2\lambda^2\mu^2+12\lambda^2\mu-6\lambda\mu^2-2\lambda^3-9\lambda^2 \\
&\quad +\mu^2-\lambda\mu+7\lambda-\mu-1\Big)+r\Big(5\lambda^2\mu-4\lambda^2\mu^2 \\
&+4\lambda\mu^2-6\lambda\mu+\mu\Big)+2\lambda^2\mu^2+4\lambda^2-6\lambda^2\mu+2\lambda\mu-2\lambda\Big]x^3+\cdots.
\end{aligned}
$$

Corollary 34. *We have*

$$C_{p;\lambda}(C_{r;\mu}, C_{u;v}) = C_{p;\lambda}$$

if we are in one of the following non-trivial cases:

$$
\begin{aligned}
&(i)\ C_{1;\lambda}(C_{1;(2\lambda-1)/\lambda}, C_{u;1}) = C_{1;\lambda}; \\
&(ii)\ C_{0;\lambda}(C_{0;(2\lambda-1)/\lambda}, C_{u;1}) = C_{0;\lambda}; \\
&(iii)\ C_0(C_{r;\mu}, C_{-r;1-\mu}) = C_0; \\
&(iv)\ C_{1/2}(C_{r;\mu}, C_{1-r;1-\mu}) = C_{1/2}; \\
&(v)\ C_1(C_{r;\mu}, C_{2-r;1-\mu}) = C_1; \\
&(vi)\ C_{0;\lambda}(C_{0;(3\lambda-1)/2\lambda}, C_0) = C_{0;\lambda}; \\
&(vii)\ C_{1;\lambda}(C_{1;(3\lambda-1)/2\lambda}, C_1) = C_{1;\lambda}; \\
&(viii)\ C_{0;1/3}(C_{r;0}, C_0) = C_{0;1/3}; \\
&(ix)\ C_{1;1/3}(C_{r;0}, C_1) = C_{1;1/3}; \\
&(x)\ C_{2;1/4}(C_{1;-1/2}, C_1) = C_{2;1/4}; \\
&(xi)\ C_{-1;1/4}(C_{0;-1/2}, C_0) = C_{-1;1/4}; \\
&(xii)\ C_{0;\lambda}(C_0, C_{0;\lambda/(2-2\lambda)}) = C_{0;\lambda}; \\
&(xiii)\ C_{1;\lambda}(C_1, C_{1;\lambda/(2-2\lambda)}) = C_{1;\lambda}; \\
&(xiv)\ C_{-1;3/4}(C_0, C_{0;3/2}) = C_{-1;3/4}; \\
&(xv)\ C_{2;3/4}(C_1, C_{1;3/2}) = C_{2;3/4}.
\end{aligned}
$$

Proof. The equivalent condition ${}^{\mathcal{C}(p;\lambda)}\mathcal{C}_{r;\mu} = \mathcal{C}_{u;v}$ gives

$${}^{\mathcal{C}(p;\lambda)}\mathcal{C}_{r;\mu}(1, 1-x) = \mathcal{C}_{u;v}(1, 1-x).$$

Equating the coefficients of $x^k, k = 1, 2, ..., 5$, were obtained the above non-trivial solutions. □

Remark 56. It is not certain that these are the only solutions of the above problem. Though the computer algebra system Maple was used, the system with six equations couldn't be solved.

Remark 57. The cases (i)–(ii), (vi)–(vii), (xii)–(xiii) and (xiv)–(xv) involving $\mathcal{C}_{1;\lambda}$ and $\mathcal{C}_{0;\lambda}$, have no similar case for $\mathcal{C}_{1/2;\lambda}$. Instead, the following results:

$${}^{\mathcal{G}(\lambda)}\mathcal{G}_{\frac{2\lambda-1}{\lambda}} = \Pi_1, \; {}^{\mathcal{G}(1/3)}\Pi_2 = \mathcal{G}, \; {}^{\mathcal{G}(\lambda)}\mathcal{G}_{\frac{3\lambda-1}{2\lambda}} = \mathcal{G}, \; {}^{\mathcal{G}(\lambda)}\mathcal{G} = \mathcal{G}_{\frac{\lambda}{2(1-\lambda)}},$$

are valid, but $\mathcal{G}_\lambda$ is not a weighted Lehmer mean.

Corollary 35. *For symmetric means we have*

$$\mathcal{C}_p(\mathcal{C}_r, \mathcal{C}_u) = \mathcal{C}_p$$

if and only if we are in the following non-trivial cases:

$$(i)\; \mathcal{C}_0(\mathcal{C}_r, \mathcal{C}_{-r}) = \mathcal{C}_0;$$
$$(ii)\; \mathcal{C}_{1/2}(\mathcal{C}_r, \mathcal{C}_{1-r}) = \mathcal{C}_{1/2};$$
$$(iii)\; \mathcal{C}_1(\mathcal{C}_r, \mathcal{C}_{2-r}) = \mathcal{C}_1.$$

Remark 58. This problem of invariance was solved in Lehmer (1971). The problem of reproducing identities,

$$\mathcal{C}_p(\mathcal{C}_r, \mathcal{C}_u) = \mathcal{C}_v,$$

was solved in Brenner and Mays (1987). The solution contains the above cases (i)–(iii) and the trivial case

$$\mathcal{C}_p(\mathcal{C}_r, \mathcal{C}_r) = \mathcal{C}_r.$$

In Costin and Toader (2013a), the problem of invariance of a weighted Lehmer mean in the family of weighted Gini means was studied. As a consequence of Theorem 58 the following was deduced:

Corollary 36. *The first terms of the series expansion of the $C_{p,\lambda}$-complementary of $\mathcal{S}_{r,r-s;\mu}$ are:*

$$
\begin{gathered}
{}^{C(p;\lambda)}\mathcal{S}_{r,r-s;\mu}(1-x) = 1 - \frac{1-2\lambda+\lambda\mu}{1-\lambda}x - \frac{\lambda(1-\mu)}{2(1-\lambda)^2}(\lambda\mu - 2r\lambda\mu + s\lambda\mu \\
+\mu - 2p\mu - s\mu + 2r\mu - 4\lambda + 4p\lambda - 2p + 2)x^2 + \frac{\lambda(1-\mu)}{6(1-\lambda)^3}(6r^2\lambda^2\mu^2 \\
+2s^2\lambda^2\mu^2 - 6sr\lambda^2\mu^2 - 18p^2\lambda^2\mu + 3s\lambda^2\mu^2 + 6p^2\lambda\mu^2 + 12pr\lambda\mu^2 - 6ps\lambda\mu^2 \\
-12r^2\lambda\mu^2 - 4s^2\lambda\mu^2 - s^2\lambda^2\mu - 9r^2\lambda^2\mu + 3rs\mu\lambda^2 + 12rs\lambda\mu^2 - 12pr\lambda^2\mu \\
+6ps\lambda^2\mu + 6p^2\lambda^3 + 3\mu^2p^2 + 18p\lambda^2\mu - 12p\lambda\mu^2 - 12pr\mu^2 - 6ps\lambda\mu \\
+6r^2\mu^2 - 6sr\mu^2 + 2s^2\mu^2 - 6p\lambda^3 + 2s^2\lambda\mu + 6qr\mu^2 - 6rs\lambda\mu + 12pr\lambda\mu \\
+6ps\mu^2 - 3p^2\lambda\mu + 15p^2\lambda^2 + 12r\lambda^2\mu - 6s\lambda^2\mu + 6r^2\lambda\mu + 12p\lambda^2\mu + \lambda^2\mu^2 \\
-27p\lambda^2 - 3p\mu^2 - s^2\mu - 3r^2\mu - 3s\mu^2 + 3p^2\mu - 15p^2\lambda + 3p\lambda\mu + 6s\lambda\mu \\
-12r\lambda\mu - 3rs\mu - 11\lambda^2\mu + 4\lambda\mu^2 + 3p^2 + 21p\lambda - 3p\mu - 2\lambda\mu + 12\lambda^2 + \mu^2 \\
-6\lambda - 3p + \mu)x^3 + \ldots
\end{gathered}
$$

This result was applied for deducing the following:

Theorem 59. *We have*

$$
{}^{C(p;\lambda)}\mathcal{S}_{r,s;\mu} = \mathcal{S}_{u,t;\nu}
$$

if ${}^{C(p;\lambda)}\mathcal{C}_{r;\mu} = \mathcal{C}_{u;\nu}$ *(thus* ${}^{C(p;\lambda)}\mathcal{S}_{r,r-1;\mu} = \mathcal{S}_{u,u-1;\nu}$*) or*

$$
\begin{gathered}
(i)\ {}^{C(1)}\mathcal{S}_{3/2,1/2} = \mathcal{S}_{u,-u} \\
(ii)\ {}^{C(1;1/5)}\mathcal{S}_{(1/2,0;-1)} = \mathcal{S}_{u,-u} \\
(iii)\ {}^{C(1;\lambda)}\mathcal{S}_{1,0;(2\lambda-1)/\lambda} = \Pi_1 \\
(iv)\ {}^{C(1)}\mathcal{S}_{r+1,r;\mu} = \mathcal{S}_{1-r,-r;1-\mu} \\
(v)\ {}^{C(1;4/5)}\mathcal{S}_{r,-r} = \mathcal{S}_{0,1/2;2} \\
(vi)\ {}^{C(1;\lambda)}\Pi_2 = \mathcal{S}_{1,0;\lambda/(1-\lambda)} \\
(vii)\ {}^{C(1/2)}\mathcal{S}_{r,s;\mu} = \mathcal{S}_{-r,-s;1-\mu} \\
(viii)\ {}^{C(0)}\mathcal{S}_{-3/2,-1/2} = \mathcal{S}_{u,-u} \\
(ix)\ {}^{C(1;1/5)}\mathcal{S}_{-1/2,0;-1} = \mathcal{S}_{u,-u} \\
(x)\ {}^{C(0;\lambda)}\mathcal{S}_{-1,0;(2\lambda-1)/\lambda} = \Pi_1 \\
(xi)\ {}^{C(0)}\mathcal{S}_{r+1,r;\mu} = \mathcal{S}_{-r-1,-r-2;1-\mu} \\
(xii)\ {}^{C(0;4/5)}\mathcal{S}_{r,-r} = \mathcal{S}_{0,-1/2;2},
\end{gathered}
$$

respectively

$$(xiii)\ {}^{C(0;\lambda)}\Pi_2 = \mathcal{S}_{0,-1;\lambda/(1-\lambda)}.$$

Remark 59. As in Remark 55 we are not sure that "if" in the enunciation of previous theorem can be replaced by "if and only if."

2.2.10 Complementariness with respect to Gini means

We pass now to the complementariness with respect to the Gini means, the case which was studied in Toader and Toader (2007a). Denote the $\mathcal{S}_{q,q-r;\nu}$-complementary of the mean M by ${}^{\mathcal{S}(q,q-r;\nu)}M$, and by ${}^{\mathcal{S}(q,q-r)}M$ if $\nu = 1/2$.

Theorem 60. *If the mean M has the series expansion*

$$M(1, 1-x) = 1 + \sum_{n=1}^{\infty} a_n x^n,$$

then ${}^{\mathcal{S}(q,q-r;\nu)}M$ *has, for* $r \neq 0$ *and* $\nu \neq 0, 1$, *the series expansion*

$${}^{\mathcal{S}(q,q-r;\nu)}M(1, 1-x) = 1 + \sum_{n=1}^{\infty} d_n x^n,$$

where

$$d_0 = 1,\ d_1 = \frac{e_1}{r},$$

$$d_n = -\frac{1}{nr} \sum_{k=0}^{n-1} [k(r+1) - n] \cdot d_k \cdot e_{n-k}\ ,\ n \geq 2,$$

with

$$e_1 = (\alpha + 1)\beta_1 - \alpha b_1, \alpha = \frac{\nu}{1-\nu},$$

$$e_n = \beta_n - \sum_{k=1}^{n-1} f_k (e_{n-k} - \beta_{n-k}) + \alpha \left[\beta_n - b_n + \sum_{k=1}^{n-1} c_k (\beta_{n-k} - b_{n-k}) \right],\ n \geq 2,$$

b_n, c_n, f_n *and* β_n *denoting the coefficients of the reduced series expansion of* M^r, M^{q-r}, N^{q-r}, *respectively* $\mathcal{S}^r_{q,q-r;\nu}$.

Proof. Denoting ${}^{\mathcal{S}(q,q-r;\nu)}M = N$, the condition $\mathcal{S}_{q,q-r;\nu}(M,N) = \mathcal{S}_{q,q-r;\nu}$ gives

$$N^{q-r}\left(N^r - \mathcal{S}^r_{q,q-r;\nu}\right) = \alpha M^{q-r}\left(\mathcal{S}^r_{q,q-r;\nu} - M^r\right)$$

Taking the values $a = 1$ and $b = 1 - x$ and denoting the coefficients of the reduced series expansion of M^r, M^{q-r}, N^r, N^{q-r} and $\mathcal{S}^r_{q,q-r;\nu}$ by b_n, c_n, e_n, f_n, respectively β_n, we get

$$\left[1+\sum_{n=1}^{\infty} f_n x^n\right]\left[\sum_{n=1}^{\infty}(e_n-\beta_n)\,x^n\right]=\alpha\left[1+\sum_{n=1}^{\infty} c_n x^n\right]\left[\sum_{n=1}^{\infty}(\beta_n-b_n)\,x^n\right]$$

This gives

$$e_1-\beta_1=\alpha\,(\beta_1-b_1),$$

$$e_n-\beta_n+\sum_{k=1}^{n-1} f_k(e_{n-k}-\beta_{n-k})=\alpha\left[\beta_n-b_n+\sum_{k=1}^{n-1} c_k(\beta_{n-k}-b_{n-k})\right],\ n\geq 2.$$

Therefore, we have a recurrence relation for e_n and using Euler's formula (2.33) we can deduce the expression of d_n. □

Corollary 37. *If the mean M has the series expansion*

$$M(1,1-x)=1+\sum_{n=0}^{\infty} a_n x^n,$$

then the first terms of the series expansion of ${}^{\mathcal{S}(q,q-r;\nu)}M$, *for* $r \neq 0$ *and* $\nu \neq 0, 1$, *are*

$$\begin{aligned}
{}^{\mathcal{S}(q,q-r;\nu)}M(1,1-x) &= 1-(1+\alpha\cdot a_1)\cdot x \\
&-\frac{\alpha}{2}\left[a_1\,(2q-r-1)\,(2+\alpha a_1+a_1)+2a_2\right]\cdot x^2 \\
-\frac{\alpha}{6(1+\alpha)}&\Big[3a_1\left(\alpha r-3rq-2\alpha q+\alpha r^2-5r\alpha q-2q+r^2+3q^2+r+5\alpha q^2\right) \\
&+3a_1^2\Big(3r^2\alpha-6q\alpha^2-10\alpha q+2r^2\alpha^2+3r\alpha^2+5r\alpha+1+2r+2\alpha+\alpha^2+r^2 \\
&-4q-12qr\alpha-9qr\alpha^2-3qr+9q^2\alpha^2+12q^2\alpha+3q^2\Big)+a_1^3\Big(2+r^2+3r \\
&+5\alpha+4\alpha^2-15qr\alpha-21qr\alpha^2-9qr\alpha^3+\alpha^3+5r^2\alpha^2+2r^2\alpha^3+9r\alpha^2+3r\alpha^3 \\
&+4r^2\alpha+9r\alpha-6q\alpha^3-18q\alpha^2-18\alpha q-3qr+15q^2\alpha+21q^2\alpha^2+9q^2\alpha^3 \\
&-6q+3q^2\Big)+6a_2(1+\alpha)\,(2q-r-1)+6a_1a_2(1+\alpha)^2\,(2q-r-1) \\
&+6a_3(1+\alpha)\Big]\cdot x^3\cdots,
\end{aligned}$$

where $\alpha = \frac{\nu}{1-\nu}$.

Remark 60. The invariance of a Gini mean with respect to Greek means was studied in Toader and Toader (2007a). A Gini mean which is invariant with respect to two Greek means is the arithmetic mean, or the geometric mean.

Another consequence was obtained in Costin and Toader (2014b).

Corollary 38. *The first terms of the series expansion of the $\mathcal{S}_{p,p-q;\lambda}$-complementary of $\mathcal{S}_{r,r-s;\mu}$ are*

$$\mathcal{S}^{(p,p-q;\lambda)}\mathcal{S}_{r,r-s;\mu}(1,1-x)=1-\frac{1-2\lambda+\lambda\mu}{1-\lambda}x$$

$$-\frac{\lambda(1-\mu)}{2(1-\lambda)^2}(\lambda\mu-2r\mu\lambda+s\mu\lambda+q\mu-2p\mu-s\mu+2r\mu-2\lambda+1+4p\lambda$$

$$-2p-2q\lambda+q)x^2+\frac{\lambda(1-\mu)}{6(1-\lambda)^3}(-1+15pq\lambda+q^2-3p\mu^2q+3\mu^2p^2$$

$$+6q\lambda^2+6q^2\lambda^2-15q\lambda^2p+3p^2-18p^2\mu\lambda^2+18p\mu q\lambda^2-6p\mu^2q\lambda$$

$$+3\mu qs\lambda-6\mu qr\lambda-6q\mu^2r\lambda+3rs\mu-5q^2\lambda-6q\lambda^3p+6p^2\mu^2\lambda+6p^2\lambda^3$$

$$-s^2\mu-3r^2\mu+6r^2\mu^2-6s\mu^2r+q^2\mu+2s^2\mu^2+\mu^2\lambda^2-3p\mu q-3q^2\mu\lambda^2$$

$$+q^2\mu^2\lambda+2\lambda+12\mu pr\lambda-6\mu ps\lambda+12p\mu^2r\lambda-6p\mu^2s\lambda+3q\mu^2s\lambda-3p^2\mu\lambda$$

$$-12p\lambda^2-15p^2\lambda+15p^2\lambda^2-12p\lambda^2r\mu+6p\lambda^2s\mu+6q\lambda^2r\mu-3q\lambda^2s\mu$$

$$-3qp-3q\lambda+6p\lambda+3rs\mu\lambda^2-6s\mu^2r\lambda^2-6rs\mu\lambda+12s\mu^2r\lambda+6r^2\mu\lambda$$

$$-12r^2\mu^2\lambda-4s^2\mu^2\lambda-s^2\mu\lambda^2-3r^2\mu\lambda^2+6r^2\mu^2\lambda^2+2s^2\mu^2\lambda^2+2s^2\mu\lambda$$

$$+6q\mu^2r-3q\mu^2s+q^2\mu^2+3\mu p^2-12p\mu^2r+6p\mu^2s-2q^2\mu\lambda+3q\mu^2\lambda$$

$$-6q\mu\lambda^2-2\lambda^2\mu-6p\mu^2\lambda+12p\mu\lambda^2+3p\mu q\lambda-6r\mu\lambda+3s\mu\lambda+6r\mu^2\lambda$$

$$-3s\mu^2\lambda+6r\mu\lambda^2-3s\mu\lambda^2-6r^2\mu\lambda^2+3s\mu^2\lambda^2)\cdot x^3+\cdots.$$

Remark 61. The next coefficient needs two pages for printing.

In the same paper, the problem of invariance in the family of weighted Gini means was also studied. The following result was found.

Theorem 61. *We have*

$$\mathcal{S}^{(p,q;\lambda)}\mathcal{S}_{r,s;\mu}=\mathcal{S}_{u,t;v}$$

if we are in a case indicated by one of the following items: Proposition 3, Remark 38, Theorem 49, Theorem 56, Theorem 57, Corollary 34, or Theorem 59.

Remark 62. In this case we have to determine sets of nine parameters which are solutions. We get only special solutions which are also solutions of one of the items indicated by the theorem. But again we cannot be sure that we got all the solutions.

Corollary 39. *We have*

$$^{\mathcal{S}(p,q)}\mathcal{S}_{r,s} = \mathcal{S}_{u,w},$$

if and only if:

$$(i) \quad ^{\mathcal{S}(p,-p)}\mathcal{S}_{r,s} = \mathcal{S}_{-r,-s};$$
$$(ii) \quad ^{\mathcal{S}(p,0)}\mathcal{S}_{r,r+p} = \mathcal{S}_{-r,p-r}.$$

This problem was also solved in Baják and Páles (2009a).

2.2.11 Complementariness with respect to Stolarsky means

The complementariness with respect to Stolarsky means was studied in Toader and Costin (2007). Denote the $\mathcal{D}_{r,s}$-complementary of the mean M by $^{\mathcal{D}(r,s)}M$.

Theorem 62. *If the mean M has the series expansion*

$$M(1,1-x) = 1 + \sum_{n=1}^{\infty} a_n x^n,$$

then $^{\mathcal{D}(r,s)}M$ has, for $r, s, r-s \neq 0$, the series expansion

$$^{\mathcal{D}(r,s)}M(1,1-x) = 1 + \sum_{n=1}^{\infty} d_n x^n,$$

with

$$d_n = -\frac{1}{nr}\sum_{k=0}^{n-1}[k(r+1)-n]\cdot d_k \cdot e_{n-k}\ ,\ n \geq 1,\ d_0 = 1,$$

where e_n can be determined from the equations

$$rf_n - se_n + sb_n - rc_n - r\cdot\sum_{k=1}^{n-1}(c_k - f_k)\,\beta_{n-k} = 0,\ n \geq 2$$

and

$$\sum_{k=0}^{n}\left[k\left(\frac{s}{r}+1\right)-n\right]\cdot e_k \cdot f_{n-k} = 0,\ n \geq 1,\ e_0 = f_0 = 1.$$

Here b_n, c_n, and β_n denote the coefficients of the reduced series expansion of M^r, M^s, respectively $\mathcal{D}_{r,s}^{r-s}$.

Proof. Denoting ${}^{\mathcal{D}(r,s)}M = N$, the condition $\mathcal{D}_{r,s}(M,N) = \mathcal{D}_{r,s}$ gives

$$s\left(M^r - N^r\right) = r\left(M^s - N^s\right)\mathcal{D}_{r,s}^{r-s}$$

Taking the values $a = 1$ and $b = 1 - x$ and denoting the coefficients of the reduced series expansion of M^r, M^s, N^r, N^s, and $\mathcal{D}_{r,s}^{r-s}$ by b_n, c_n, e_n, f_n, respectively β_n, we get

$$s\left[\sum_{n=1}^{\infty}(b_n - e_n)\,x^n\right] = r\left[\sum_{n=1}^{\infty}(c_n - f_n)\,x^n\right]\left[1 + \sum_{n=1}^{\infty}\beta_n x^n\right],$$

or

$$\sum_{n=1}^{\infty}\left[s\,(b_n - e_n) - r\,(c_n - f_n)\right]\cdot x^n = r\cdot\sum_{n=2}^{\infty}\left[\sum_{k=1}^{n-1}(c_k - f_k)\,\beta_{n-k}\right]\cdot x^n.$$

This gives

$$s\,(b_1 - e_1) - r\,(c_1 - f_1) = 0$$

and

$$s\,(b_n - e_n) - r\,(c_n - f_n) = r\cdot\sum_{k=1}^{n-1}(c_k - f_k)\,\beta_{n-k},\ n \geq 2.$$

Write $M^s = (M^r)^{s/r}$ and use Euler's formula (2.33) to deduce the announced expressions. □

Corollary 40. *If the mean M has the series expansion*

$$M(1, 1-x) = 1 + \sum_{n=0}^{\infty} a_n x^n,$$

then the first terms of the series expansion of ${}^{\mathcal{D}(r,s)}M$, for $r, s, r - s \neq 0$, are

$$ {}^{\mathcal{D}(r,s)}M(1, 1-x) = 1 - (a_1 + 1)\cdot x - \frac{1}{3}[a_1(a_1+1)(s+r-3) + 3a_2]\cdot x^2$$

$$+\frac{1}{18}[2a_1^3(6s - s^2 - 2sr - r^2 + 6r - 9) + 3a_1^2(5s - s^2 - 2sr - r^2 - 6 + 5r)$$

$$+a_1(3s - s^2 - 2sr - r^2 + 3r) + 6a_2(3 - s - r)(1 + 2a_1) - 18a_3]\cdot x^3 + \frac{1}{4320}$$

$$\cdot[8a_1^4(540 - 525r - 525s + 360rs + 180r^2 + 180s^2 - 69r^2s - 69rs^2 - 19r^3$$

$$-19s^3) + 8a_1^3(540 - 690r - 690s + 600sr + 300r^2 + 300s^2 - 138rs^2$$

$$-138r^2s - 38r^3 - 38s^3) + 8a_1^2(240sr - 135r - 135s + 120s^2 + 120r^2$$
$$-81rs^2 - 81r^2s - 21r^3 - 21s^3) + 16a_1(15r + 15s - 6rs^2 - 6r^2s - r^3 - s^3)$$
$$+1440a_1^2a_2(6r + 6s - 9 - 2rs - r^2 - s^2) + 1440a_1a_2(5r + 5s - 6 - s^2 - r^2$$
$$-2rs) + 2880a_1a_3(3 - r - s) + 240a_2(3r + 3s - s^2 - r^2 - 2rs)$$
$$+1440a_3(3 - s - r) - 4320a_4 + 1440a_2^2(3 - s - r) + 3(225 - 70s - 70r$$
$$-10sr - 5r^2 - 5s^2 + 2r^2s + 2rs^2 + 2r^3 + 2s^3)] \cdot x^4 + \cdots$$

The following consequences were proved in Toader and Toader (2007b).

Corollary 41. *The first terms of the series expansion of the $\mathcal{D}_{r,s}$-com plementary of $\mathcal{P}_{q,t}$ are*

$$^{\mathcal{D}(r,s)}\mathcal{P}_{q,t}(1, 1-x) = 1 - t \cdot x + t(1-t)(2r + 2s - 3q - 3) \cdot \frac{x^2}{6}$$
$$-t(t-1)(6tq^2 + 2s^2t + 2r^2t + 4srt - 6rtq - 6stq - r^2 - s^2 - 3q^2$$
$$-6st - 6rt + 9tq + 3sq + 3rq - 2sr + 6r + 6s - 9q + 3t - 6) \cdot \frac{x^3}{18}$$
$$+(1080t^4q^3 + 720s^2t^4q + 720r^2t^4q - 1320rt^4q^2 + 1440srt^4q - 552s^2rt^4$$
$$-1320st^4q^2 - 152s^3t^4 - 552sr^2t^4 - 152r^3t^4 + 1980t^4q^2 + 720r^2t^4$$
$$-2160rt^4q - 2160st^4q - 1440r^2t^3q + 2640rt^3q^2 - 2880srt^3q + 304s^3t^3$$
$$-1440s^2t^3q + 720s^2t^4 + 304r^3t^3 + 1104s^2rt^3 - 2160t^3q^3 - 648sr^2t^2$$
$$+1440srt^4 + 1104sr^2t^3 + 2640st^3q^2 + 840r^2t^2q + 840s^2t^2q + 1080t^4q$$
$$-120s^2tq - 1560rt^2q^2 + 1260t^2q^3 - 720rt^4 - 720st^4 - 1560st^2q^2$$
$$-168s^3t^2 + 1680srt^2q - 5400t^3q^2 - 1920r^2t^3 + 5760rt^3q + 5760st^3q$$
$$-3840srt^3 - 1920s^2t^3 - 648s^2rt^2 + 3120srt^2 - 4680st^2q + 240rtq^2$$
$$-240srtq - 120r^2tq - 180tq^3 + 96sr^2t + 180t^4 - 168r^3t^2 + 16r^3t$$
$$+240stq^2 + 16s^3t - 4680rt^2q + 4500t^2q^2 - 4320t^3q + 1560r^2t^2 + 2880rt^3$$
$$+2880st^3 + 1560s^2t^2 + 96s^2rt + 6sr^2 + 6s^2r + 6s^3 + 6r^3 + 5220t^2q$$
$$-3480st^2 - 3480rt^2 - 1080tq^2 - 360r^2t + 1080rtq + 1080stq$$
$$-1080t^3 - 720srt - 360s^2t - 15r^2 - 15s^2 - 30sr - 1980tq + 1980t^2$$
$$+1320st + 1320rt - 210s - 210r - 1080t + 675) \cdot \frac{x^4}{4320} + \cdots$$

Using it we can prove results of the following type.

Theorem 63. *The relation*

$$^{\mathcal{D}(r,s)}\mathcal{P}_{p,\mu} = \mathcal{P}_{q,\nu}, \; rs\,(r-s) \neq 0,$$

holds in and only in the following cases:

$$\begin{aligned}
&(i)\ ^{\mathcal{D}(r,s)}\mathcal{P}_{p,0} = \mathcal{P}_{q,1};\\
&(ii)\ ^{\mathcal{D}(r,s)}\mathcal{P}_{p,1} = \mathcal{P}_{q,0};\\
&(iii)\ ^{\mathcal{D}(r,-r)}\mathcal{P}_{p,\mu} = \mathcal{P}_{-p,1-\mu};\ or\\
&(iv)\ ^{\mathcal{D}(r,2r)}\mathcal{P}_{r,\mu} = \mathcal{P}_{r,1-\mu}.
\end{aligned}$$

Proof. Consider the reduced series expansion on both sides of the desired equality. The coefficients of x are the same if and only if

$$\nu = 1 - \mu.$$

Equating the coefficients of x^2 one obtains:

$$\mu\,(1-\mu)\,(2r + 2s - 3p - 3q) = 0.$$

This happens if:
1. $\mu = 0$, giving (i);
2. $\mu = 1$, thus (ii); or
3. $2r + 2s - 3p - 3q = 0$. Equating also the coefficients of x^3, in this case follows

$$(1 - 2\mu)\,(r + s)\,(r - 3p + s) = 0.$$

It splits into three sub-cases:
3.1. $r = -s$ which gives $q = -p$, thus (iii);
3.2. $r - 3p + s = 0$, which need to be combined with the equality of the coefficients of x^4 giving

$$(2r - s)\,(r - 2s)\left(2\mu^2 - 2\mu + 1\right) = 0,$$

which leads to (iv); or
3.3. $\mu = \frac{1}{2}$ which, combined with the equality of the coefficients of x^5 and x^6, give only special cases of (iv).

We have no more cases. On the other hand, writing these cases as

$$(i)\ ^{\mathcal{D}(r,s)}\Pi_2 = \Pi_1;\ (ii)\ ^{\mathcal{D}(r,s)}\Pi_1 = \Pi_2;\ (iii)\ ^{\mathcal{G}}\mathcal{P}_{p,\mu} = \mathcal{P}_{-p,1-\mu};$$
$$(iv)\ ^{\mathcal{P}(r)}\mathcal{P}_{r,\mu} = \mathcal{P}_{r,1-\mu};$$

we can directly verify that they are true. $\square$

The invariance in the family of Stolarsky means was studied in Toader and Costin (2007).

Corollary 42. *The first terms of the series expansion of the $\mathcal{D}_{r,s}$-complementary of $\mathcal{D}_{p,q}$ are*

$$
\begin{aligned}
{}^{\mathcal{D}(r,s)}\mathcal{D}_{p,q}(1,1-x) &= 1-\frac{x}{2}+(2r+2s-p-q-3)\cdot\frac{x^2}{24}\\
&+(2r+2s-p-q-3)\cdot\frac{x^3}{48}+(2r^3+2s^3+6p^3+6q^3+6p^2q+6pq^2\\
&-10rp^2+20s^2p+20r^2p-10sp^2-18rs^2-18sr^2+20s^2q-10sq^2+20r^2q\\
&-10rq^2+40srp+40srq-20rpq-20spq+15p^2+15q^2+30pq\\
&-60rq-60rp-60sq-60sp-420s-420r-210p-210q+2025)\\
&\cdot\frac{x^4}{17280}+\cdots.
\end{aligned}
$$

Corollary 43. *There are no values $\rho, q, m, n \in \mathbb{R}$ and $t \in [0,1]$ such that*

$$
{}^{\mathcal{D}(r,s)}\mathcal{D}_{p,q}=\mathcal{S}_{m,m-n,t}
$$

for some $r, s, r-s \neq 0$.

Proof. The coefficients of x respectively x^2 of the reduced series expansion of the two members are equal if and only if

$$
t=\frac{1}{2}, n=\frac{6m-2r-2s+p+q}{3},
$$

but the equality of the coefficients of x^3 holds for no value of the parameters. □

Corollary 44. *If*

$$
{}^{\mathcal{D}(r,s)}\mathcal{D}_{p,q}=\mathcal{D}_{u,v},\ \textit{for } rspquv\,(r-s)\,(p-q)\,(u-v)\neq 0
$$

then

(i) ${}^{\mathcal{D}(r,-r)}\mathcal{D}_{p,-p}=\mathcal{D}_{u,-u}$;

(ii) ${}^{\mathcal{D}(r,-r)}\mathcal{D}_{p,q}=\mathcal{D}_{-p,-q}$;

(iii) ${}^{\mathcal{D}(r,-r)}\mathcal{D}_{p,q}=\mathcal{D}_{-q,-p}$; *or*

(iv) $v=2r+2s-p-q-u,$
$13(p+q-r-s)^2(r+s)+24u(p+q)(r+s)-6u(p+q)^2$
$+6u^2(2r+2s-p-q)-24u(r+s)^2-6pq(p+q)$
$+12sr(s+r)=0,$

$$946pq(r^3+s^3)-1089pq(r^2+s^2)(p+q)+464pq(r+s)(p^2+q^2)$$
$$+834q^2p^2(r+s)-3066s^2r^2(p+q)+40(s^5+r^5)-10(p^5+q^5)$$
$$+1863sr(s+r)(p^2+q^2)+1480s^2r^2(s+r)$$
$$-267(r^2+s^2)(p^3+q^3)-726sr(p^3+q^3)-38pq(p^3+q^3)$$
$$+560rs(r^3+s^3)-208q^2p^2(p+q)+3726rspq(r+s)$$
$$+473(p^2+q^2)(r^3+s^3)-283(r^4+s^4)(p+q)$$
$$-1744rs(r^2+s^2)(p+q)-2466rspq(p+q)$$
$$+47(r+s)(p^4+q^4)=0.$$

Proof. The coefficients of x in the series expansion of the two members are equal. The equality of the coefficients of x^2 or x^3 leads to

$$v=2r+2s-p-q-u.$$

Similarly, equating the coefficients of x^4 or x^5 gives

$$13(p+q-r-s)^2(r+s)+24u(p+q)(r+s)-6u(p+q)^2+$$
$$+6u^2(2r+2s-p-q)-24u(r+s)^2-6pq(p+q)+12sr(s+r)=0.$$

Then the coefficients of x^6 or x^7 imply

$$(r+s)[946pq(r^3+s^3)-1089pq(r^2+s^2)(p+q)+464pq(r+s)(p^2+q^2)$$
$$+834q^2p^2(r+s)-1744rs(r^2+s^2)(p+q)-3066s^2r^2(p+q)$$
$$+1480s^2r^2(s+r)+1863sr(s+r)\left(p^2+q^2\right)-10\left(p^5+q^5\right)+40\left(s^5+r^5\right)$$
$$-38pq\left(p^3+q^3\right)+473\left(p^2+q^2\right)\left(r^3+s^3\right)-283\left(r^4+s^4\right)(p+q)$$
$$-726sr\left(p^3+q^3\right)+3726rspq(r+s)+560rs\left(r^3+s^3\right)$$
$$-267\left(r^2+s^2\right)\left(p^3+q^3\right)-2466rspq(p+q)+47(r+s)\left(p^4+q^4\right)$$
$$-208q^2p^2(p+q)\Big]=0.$$

1. If $s=-r$, then

$$(p+u)(q+u)(p+q)=0.$$

This implies:
1.1. $q=-p$ and $v=-u$, so (i);
1.2. $u=-p$ and $v=-q$, thus (ii); or
1.3. $u=-q$ and $v=-p$, giving (iii).
2. If $s\neq -r$, then we get (iv). □

Corollary 45. *We have*

$$^{\mathcal{G}}\mathcal{D}_{p,q} = \mathcal{D}_{u,v}, \text{ for } pquv\,(p-q)\,(u-v) \neq 0$$

if and only if

$$\begin{aligned} &(i) \quad {}^{\mathcal{G}}\mathcal{D}_{p,-p} = \mathcal{D}_{u,-u}; \\ &(ii) \quad {}^{\mathcal{G}}\mathcal{D}_{p,q} = \mathcal{D}_{-p,-q}; \text{ or} \\ &(iii) \quad {}^{\mathcal{G}}\mathcal{D}_{p,q} = \mathcal{D}_{-q,-p}. \end{aligned}$$

Remark 63. This result was proved by another method, even for $pquv\ (p-q)(u-v)=0$, in Błasińska-Lesk et al. (2003).

Corollary 46. *If*

$$^{\mathcal{D}(r,s)}\mathcal{D}_{p,s} = \mathcal{D}_{u,s}, \text{ for } rspu\,(r-s)\,(p-s)\,(u-s) \neq 0$$

then

$$s = -13r,\ p = r\left(1+3\sqrt{10}\right),\ u = r\left(1-3\sqrt{10}\right).$$

Proof. Taking $s=q=v$ we get the above relations among the parameters. □

Corollary 47. *There is no non-trivial case such that*

$$^{\mathcal{D}(r,1)}\mathcal{D}_{p,1} = \mathcal{D}_{u,1}, \text{ for } rpu\,(r-1)\,(p-1)\,(u-1) \neq 0.$$

Proof. Taking $r=-1/13$ in the previous corollary, for $a=2500$ we have

$$\mathcal{D}_{r,1}\left(\mathcal{D}_{p,1}(a,1), \mathcal{D}_{u,1}(a,1)\right) - \mathcal{D}_{r,1}(a,1) \geq 1.$$

□

Remark 64. This result was also proved in Liu (2004).

Remark 65. To find all the solutions of the invariance problem, we have to solve a system with at least six equations. Knowing three equations, we have to add the equality of the coefficients of x^8, x^{10} and x^{12}. As before, the equality of the coefficients of x^9 and x^{11} gives the same equations as the coefficients of x^8 and x^{10}. These equations are too complicated (for instance, the last equation can be written on 30 pages). On a usual good performance computer, as that described in Heck (2003), Maple cannot solve this system, working 24 hours. As we showed, we obtained only some special results. However, the problem was completely solved otherwise in Baják and Páles (2012). As it is proved there, the only non-trivial solution of the invariance problem in the class of Stolarsky means is

$$\mathcal{G}\left(\mathcal{D}_{p,q}, \mathcal{D}_{-q,-p}\right) = \mathcal{G}.$$

2.2.12 Complementariness with respect to extended logarithmic means

The problem of complementariness with respect to the logarithmic mean was treated in Costin (2008), Costin and Toader (2009) and Costin (2010). It was generalized to extended logarithmic means in Toader et al. (2012).

Denote the $\mathcal{L}_p$-complementary of the mean M by ${}^{\mathcal{L}(p)}M$.

Lemma 14. *If the mean M^p has the series expansion*

$$M^p(1,1-x)=1+\sum_{n=1}^{\infty}a_n x^n,$$

then the first coefficients of the series expansion of $\left({}^{\mathcal{L}(p)}M\right)^p$ are

$$\left({}^{\mathcal{L}(p)}M\right)^p(1,1-x)=1-(a_1+p)x+\frac{1}{6}\left(4a_1^2+4pa_1-6a_2+3p^2-3p\right)x^2$$
$$+\frac{1}{18}[-8a_1^3-6pa_1^2+2pa_1(3-2p)+24a_1a_2-18a_3+12pa_2-3p(p^2-3p$$
$$+2)]x^3+\frac{1}{1080}\Big[352a_1^4+224pa_1^3+36pa_1^2(3p-5)-1440a_1^2a_2-720pa_1a_2$$
$$+1440a_1a_3+8pa_1(7p^2-30p+30)+720a_2^2-120pa_2(2p-3)+720pa_3$$
$$-1080a_4+45p\,(p^3-6p^2+11p-6)\Big]x^4+\ldots$$

Proof. Denoting ${}^{\mathcal{L}(p)}M=N$ we have the condition $\mathcal{L}_p\,(M,N)=\mathcal{L}_p$, thus

$$\frac{M^p-N^p}{\mathcal{L}_p^p}=\ln M^p-\ln N^p.$$

For

$$N^p(1,1-x)=1+\sum_{n=1}^{\infty}b_n x^n,$$

we have

$$\frac{\sum_{n=0}^{\infty}a_n x^n-\sum_{n=0}^{\infty}b_n x^n}{\left(\sum_{n=0}^{\infty}h_n x^n\right)^p}=\ln\left(\sum_{n=0}^{\infty}a_n x^n\right)-\ln\left(\sum_{n=0}^{\infty}b_n x^n\right). \tag{2.45}$$

Putting

$$\frac{\sum_{n=0}^{\infty} a_n x^n - \sum_{n=0}^{\infty} b_n x^n}{\left(\sum_{n=0}^{\infty} h_n x^n\right)^p} = \sum_{n=0}^{\infty} c_n x^n$$

and using the denotations from (2.34), we obtain

$$\sum_{n=1}^{\infty} (a_n - b_n) x^n = \left(\sum_{n=0}^{\infty} l_n x^n\right)\left(\sum_{n=0}^{\infty} c_n x^n\right),$$

thus

$$c_0 = 0, c_1 = a_1 - b_1, c_n = a_n - b_n - \sum_{k=1}^{n-1} l_k c_{n-k}, n > 1.$$

The equality (2.45) becomes

$$\sum_{n=1}^{\infty} c_n x^n = \ln\left(\sum_{n=0}^{\infty} a_n x^n\right) - \ln\left(\sum_{n=0}^{\infty} b_n x^n\right)$$

and we deduce

$$\sum_{n=1}^{\infty} n c_n x^{n-1} = \frac{\sum_{n=1}^{\infty} n a_n x^{n-1}}{\sum_{n=0}^{\infty} a_n x^n} - \frac{\sum_{n=1}^{\infty} n b_n x^{n-1}}{\sum_{n=0}^{\infty} b_n x^n},$$

thus

$$\left(\sum_{n=0}^{\infty} a_n x^n\right)\left(\sum_{n=0}^{\infty} b_n x^n\right)\left[\sum_{n=0}^{\infty} (n+1) c_{n+1} x^n\right]$$

$$= \left(\sum_{n=0}^{\infty} b_n x^n\right)\left[\sum_{n=0}^{\infty} (n+1) a_{n+1} x^n\right] - \left(\sum_{n=0}^{\infty} a_n x^n\right)\left[\sum_{n=0}^{\infty} (n+1) b_{n+1} x^n\right],$$

or

$$\sum_{n=0}^{\infty} \left[d_0 (n+1) c_{n+1} + d_1 n c_n + \ldots + d_n c_1\right] x^n = \sum_{n=0}^{\infty} \left[b_0 (n+1) a_{n+1}\right.$$

$$+b_1 n a_n + ... + b_n a_1 - a_0 (n+1) b_{n+1} - a_1 n b_n - ... - a_n b_1] x^n,$$

where $d_n = a_0 b_n + a_1 b_{n-1} + ... + a_n b_0$. The equality of the coefficients of the same power of x allows the step-by-step determination of b_n. □

Corollary 48. *If the mean M has the series expansion*

$$M(1, 1-x) = 1 + \sum_{n=1}^{\infty} \alpha_n x^n,$$

then the first coefficients of the series expansion of $\left({}^{\mathcal{L}(p)}M\right)^p$ are

$$\begin{aligned}({}^{\mathcal{L}(p)}M)^p(1,1-x) = {} & 1 - p(\alpha_1+1)x + \frac{p}{6}[(\alpha_1^2+4\alpha_1+3)p + 3\alpha_1^2 - 6\alpha_2 - 3]x^2 \\ & + \frac{p}{18}\left[\left(\alpha_1^3 - 4\alpha_1 - 3\right)p^2 - 3\left(\alpha_1^3 + 2\alpha_1^2 - 2\alpha_1\alpha_2 - 2\alpha_1 - 4\alpha_2 - 3\right)p \right. \\ & \left. + 6\left(3\alpha_1\alpha_2 - \alpha_1^3 - 3\alpha_3 - 1\right)\right]x^3 + \frac{p}{1080}\left[\left(7\alpha_1^4 - 16\alpha_1^3 - 12\alpha_1^2 + 56\alpha_1 \right.\right. \\ & \left. + 45\right)p^3 + \left(180\alpha_1^2\alpha_2 - 90\alpha_1^4 + 120\alpha_1^2 - 240\alpha_1 - 240\alpha_2 - 270\right)p^2 \\ & + \left(165\alpha_1^4 + 240\alpha_1^3 - 180\alpha_1^2 - 540\alpha_1^2\alpha_2 + 240\alpha_1 - 720\alpha_1\alpha_2 \right. \\ & \left. + 360\alpha_1\alpha_3 + 360\alpha_2 + 495 + 180\alpha_2^2 + 720\alpha_3\right)p + 1080\alpha_1\alpha_3 \\ & \left. + 270\alpha_1^4 + 540\alpha_2^2 - 1080\alpha_4 - 270 - 1080\alpha_1^2\alpha_2\right]x^4 + ...\end{aligned}$$

Proof. Given the coefficients α_n we deduce the coefficients a_n by Euler's formula:

$$a_n = \frac{1}{n}\sum_{k=1}^{n}\left[k(p+1) - n\right]\alpha_k a_{n-k}, n \geq 1,$$

obtaining the result. □

Theorem 64. *If the mean M has the series expansion*

$$M(1, 1-x) = 1 + \sum_{n=1}^{\infty} \alpha_n x^n,$$

then the first coefficients of the series expansion of ${}^{\mathcal{L}(p)}M$ are

$$\begin{aligned}{}^{\mathcal{L}(p)}M(1, 1-x) = {} & 1 - (\alpha_1+1)x + \frac{1}{3}\left[\alpha_1(\alpha_1+1)(3-p) - 3\alpha_2\right]x^2 \\ & + \frac{1}{18}\left[-2(p-3)^2\alpha_1^3 + 3\left(5p - 9 - p^2\right)\alpha_1^2 + p(3-p)\alpha_1 + 12(3-p)\alpha_1\alpha_2\right.\end{aligned}$$

$$-18\alpha_3+6(3-p)\alpha_2\big]x^3+\frac{1}{540}\Big[\left(540-525p+180p^2-19p^3\right)\alpha_1^4$$
$$+\left(540-690p+300p^2-38p^3\right)\alpha_1^3+p\left(-135+120p-21p^2\right)\alpha_1^2$$
$$+p\left(30-2p^2\right)\alpha_1+180(3-p)\alpha_2^2+30p(3-p)\alpha_2$$
$$-180(3-p)^2\alpha_1^2\alpha_2-180(3-p)(2-p)\alpha_1\alpha_2$$
$$+360(3-p)\alpha_1\alpha_3+180(3-p)\alpha_3-540\alpha_4\big]x^4+\ldots$$

Proof. Given the coefficients b_n we deduce the coefficients β_n of ${}^{\mathcal{L}(p)}M(1, 1-x)$ by Euler's formula:

$$\beta_n=\frac{1}{n}\sum_{k=1}^{n}\left[k\left(\frac{1}{p}+1\right)-n\right]b_k\beta_{n-k}, n\geq 1.$$ □

Using the previous result, in Toader and Toader (2010) the following results were deduced.

Theorem 65. *The first terms of the series expansion of the $\mathcal{L}_p$-complementary of $\mathcal{S}_{q,q-r;\nu}$ are*

$${}^{\mathcal{L}(p)}\mathcal{S}_{q,q-r;\nu}(1,1-x)=1-\nu x+\nu(1-\nu)(2p+3r-6q-3)\cdot\frac{x^2}{6}$$
$$+\nu(1-\nu)\Big(2p^2\nu-6p\nu-12\nu pq+6\nu pr+6\nu r^2-18\nu rq-9r\nu+18\nu q^2$$
$$+18q\nu-p^2+6p+6pq-3pr+3\nu+9r-3r^2-18q+9rq-9q^2-6\Big)\cdot\frac{x^3}{18}$$
$$+\nu(1-\nu)\Big(-270+180p^2r\nu^2-180p^2r\nu-360p^2q\nu^2+1080pq\nu^2$$
$$-540pr\nu^2+360p^2q\nu+495r-990q-60p^2q-1620rq^2\nu-180q^3-810q^2$$
$$-45\nu^2+30p^2r+1080pq^2\nu^2-1080pq^2\nu+330pr^2\nu^2-330pr^2\nu-90p^2$$
$$-38p^3\nu-180pqr+270rq^2+60pr^2+180pq^2+38p^3\nu^2-1080pqr\nu^2$$
$$+1080pqr\nu+4p^3+1080r^2q\nu+300p^2\nu-180p^2\nu^2-270r^2+270r\nu^2$$
$$-270r^3\nu-540q\nu^2+45r^3+810rq-1080q^3\nu^2-1620q^2\nu^2-540p\nu$$
$$-270pr+270r^3\nu^2+1620qr\nu^2-1080r^2q\nu^2+540pq-495r^2\nu^2$$
$$+1620rq^2\nu^2+1080q^3\nu+1620q\nu-810r\nu+180p\nu^2+330p+225\nu$$
$$-180r^2q-1800pq\nu+900pr\nu-2700rq\nu+2700q^2\nu+855r^2\nu\Big)$$
$$\cdot\frac{x^4}{1080}+\cdots.$$

Theorem 66. *If*

$$^{\mathcal{L}(p)}\mathcal{S}_{q,q-r;\nu} = \mathcal{S}_{t,t-s;\mu}, \ \textit{for } rt \neq 0$$

then we can get non-trivial cases only if

$$\nu = \mu = \frac{1}{2}.$$

Proof. Equating the coefficients of x we have

$$\mu = 1 - \nu.$$

Passing to the coefficients of x^2, we get the trivial cases $\nu = 0$ and $\nu = 1$, or the condition

$$2p - 6q + 3t + 3r - 6s = 0.$$

This condition, for the coefficients of x^3, can be true in one of the following three cases: $\nu = \mu = \frac{1}{2}$, $r = t$ or $r = -t$. Equating the coefficients of $x^k, k = 1, 2, 3, 4$, we get:

$$1) \ \nu = \mu = \frac{1}{2}, q = -s;$$

$$2) \ \nu = \mu = \frac{1}{2}, r = t;$$

or

$$3) \ \nu = \mu = \frac{1}{2}, r = -t.$$

□

Corollary 49. *If*

$$^{\mathcal{L}(p)}\mathcal{S}_{q,q-r} = \mathcal{S}_{t,t-s}, \ \textit{for } rt \neq 0,$$

then we have at most the following cases:

$$t = r, \ p/r = -1.860..., q/r = -0.413..., s/r = -0.033...;$$

$$t = -r, \ p/r = -1.860..., q/r = 0.413..., s/r = -1.033...$$

or

$$q = -s, \ p/r = 1.635..., q/r = 1.081..., t/r = -2.090...$$

Proof. Equating the coefficients of $x^k, k = 1, 2, ..., 6$, we get the above solutions. □

Corollary 50. *If*

$$^{\mathcal{L}(p)}\mathcal{C}_q = \mathcal{C}_t,$$

then we can have the following cases:

$$(i)\ p = 1.629...,\ q = 1.081...,\ t = 2.076...;$$
$$(ii)\ p = 1.802...,\ q = 0.582...,\ t = 0.963...;$$
$$(iii)\ p = 2.177...,\ q = 1.091...,\ t = 1.731...;$$
$$(iv)\ p = -1.490...,\ q = -0.069...,\ t = 1.854...$$

Corollary 51. *If*

$$^{\mathcal{L}(p)}\mathcal{P}_q = \mathcal{S}_{t,t-s},$$

then we can have the following cases:

$$(i)\ p/t = 1.575...,\ q/t = 0.335...,\ s/t = 0.857...$$

or

$$(ii)\ p/t = -2.058...,\ q/t = 0.350...,\ s/t = -0.361...$$

Corollary 52. *If*

$$^{\mathcal{L}(p)}\mathcal{P}_q = \mathcal{C}_t,$$

then we can have at most the following cases:

$$(i)\ p = 1.787...,\ q = 2.718...,\ t = 3.526...;$$
$$(ii)\ p = 1.836...,\ q = 0.390...,\ t = 1.166...;$$
$$(iii)\ p = 3.739...,\ q = 2.082...,\ t = 1.589...;$$
$$(iv)\ p = -0.707...,\ q = -1.075...,\ t = 1.395...;$$
$$(v)\ p = -6.340...,\ q = -3.531...,\ t = 2.695...$$

or

$$(vi)\ p = -11.056...,\ q = -2.351...,\ t = 7.019...$$

Remark 66. Unfortunately none of the above determined possible solutions can be verified. To find the right solutions of the above problem, we have to add more equations to the system. These equations are quite complex and they are unsolvable by Maple 8 running on a usual computer.

Returning to the results of Toader et al. (2012), we have:

Theorem 67. *The first terms of the series expansion of the $\mathcal{L}_p$-complementary of $\mathcal{L}_q$ are*

$$\begin{aligned} {}^{\mathcal{L}(p)}\mathcal{L}_q(1,1-x) &= 1-\frac{x}{2}+(2p+q-9)\cdot\frac{x^2}{48}(2+x) \\ &-\Big[22p^3+20(q-9)p^2+10\left(q^2-18q+30\right)p \\ &-3\left(2q^3+5q^2-190q+135\right)\Big]\cdot\frac{x^4}{17280}+\cdots. \end{aligned}$$

Corollary 53. *There is no case such that*

$$^{\mathcal{L}(p)}\mathcal{L}_q=\mathcal{L}_r,\ pqr\neq 0.$$

Proof. Equating the coefficients of x^2 we get the condition

$$2p+q+r=12.$$

But then, equating the coefficients of x^k, $k=0,1,...,5$, we get no solution. □

Remark 67. If we take $\mathcal{L}_0(a,b)=\sqrt{ab}$, we deduce that the problem of invariance in the class of extended logarithmic means has the only solution

$$\mathcal{L}_0\left(\mathcal{L}_q,\mathcal{L}_{-q}\right)=\mathcal{L}_0.$$

More generally, in Costin (2010) the following was proved:

Theorem 68. *The complementary ${}^{\mathcal{L}}D_{r,s}$ of a Stolarsky mean $D_{r,s}$ with respect to the logarithmic mean $\mathcal{L}$ is again a Stolarsky mean $\mathcal{D}_{u,v}$ if and only if $\mathcal{D}_{r,s}=\mathcal{D}_{u,v}=\mathcal{L}$.*

2.2.13 Complementariness with respect to the identric mean

Some general necessary conditions for a mean N to be the complementary of another mean M with respect to a given mean P were determined in Costin and Toader (2008a).

Theorem 69. *If the mean P has continuous partial derivatives of any order, with $P_b(1,1)\neq 0$ and the mean M has the series expansion*

$$M(1,1-x)=1+c_1x+c_2x^2+c_3x^3+\cdots,$$

then the series expansion of the complementary of M with respect to P, thus

$$N(1,1-x)=1+d_1x+d_2x^2+d_3x^3+\cdots,$$

has the following first coefficients

$$d_1 = -\frac{1}{p_b}\left(p_a c_1 + p_b\right),$$

$$d_2 = -\frac{1}{2p_b}\left[p_{a^2}c_1^2 + 2p_{ab}c_1 d_1 + p_{b^2}\left(d_1^2 - 1\right) + 2p_a c_2\right]$$

and

$$d_3 = -\frac{1}{6p_b}\left[p_{a^3}c_1^3 + 3p_{a^2b}c_1^2 d_1 + 3p_{ab^2}c_1 d_1^2 + p_{b^3}\left(d_1^3 + 1\right)\right.$$
$$\left. + 6\left(p_{a^2}c_1c_2 + p_{ab}c_1d_2 + p_{ab}c_2d_1 + p_{b^2}d_1d_2 + p_a c_3\right)\right],$$

where

$$p_{a^i b^j} = P_{a^i b^j}(1,1),\ i, j \geq 0.$$

Proof. If we denote

$$f(x) = P(1, 1-x),$$
$$g(x) = M(1, 1-x)$$

and

$$h(x) = N(1, 1-x),$$

we have the condition

$$f(x) = P(g(x), h(x)).$$

Therefore we get successively

$$f'(x) = P_a(g(x), h(x))g'(x) + P_b(g(x), h(x))h'(x),$$
$$f''(x) = P_{a^2}(g(x), h(x))g'^2(x) + 2P_{ab}(g(x), h(x))g'(x)h'(x)$$
$$+ P_{b^2}(g(x), h(x))h'^2(x) + P_a(g(x), h(x))g''(x) + P_b(g(x), h(x))h''(x),$$

respectively

$$f'''(x) = P_{a^3}(g(x), h(x))g'^3(x) + 3P_{a^2b}(g(x), h(x))g'^2(x)h'(x)$$
$$+ 3P_{ab^2}(g(x), h(x))g'(x)h'^2(x) + P_{b^3}(g(x), h(x))h'^3(x)$$
$$+ 3[P_{a^2}(g(x), h(x))g'(x)g''(x)$$
$$+ P_{ab}(g(x), h(x))g'(x)h''(x) + P_{ab}(g(x), h(x))g''(x)h'(x)$$
$$+ P_{b^2}(g(x), h(x))h'(x)h''(x)]$$

$$+P_a(g(x),h(x))g'''(x)+P_b(g(x),h(x))h'''(x).$$

Taking $x=0$, as

$$f(x)=f(0)+f'(0)x+\frac{f''(0)}{2}x^2+\frac{f'''(0)}{6}x^3+\cdots,$$

but also

$$f(x)=P(1,1-x)=1-p_bx+\frac{p_{b^2}}{2}x^2-\frac{p_{b^3}}{6}x^3+\cdots,$$

we obtain the above coefficients. □

Corollary 54. *If the symmetric mean P has continuous partial derivatives up to order 3 and the mean M has the series expansion*

$$M(1,1-x)=1+c_1x+c_2x^2+c_3x^3+\cdots,$$

then the first coefficients of the series expansion of the complementary of M with respect to P are

$$N(1,1-x)=1-(c_1+1)x-[4\alpha c_1(c_1+1)+c_2]x^2$$

$$-[24\alpha^2c_1(c_1+1)(2c_1+1)+12\alpha c_2(2c_1+1)-4\beta c_1(c_1+1)+3c_3]\frac{x^3}{3}+\cdots,$$

where

$$\alpha=P_{a^2}(1,1) \text{ and } \beta=P_{a^3}(1,1).$$

Proof. As the mean P is symmetric, we have (2.27), (2.29) and (2.31), giving the above coefficients. □

Theorem 70. *If the mean M has the series expansion*

$$M(1,1-x)=1+c_1x+c_2x^2+c_3x^3+\cdots,$$

then the series expansion of the complementary of M, with respect to $\mathcal{I}$, has the following first coefficients

$$^{\mathcal{I}}M(1,1-x)=1-(c_1+1)x-\frac{1}{3}\left(3c_2-c_1^2-c_1\right)x^2$$
$$-\frac{1}{18}\left(18c_3-12c_1c_2-6c_2+2c_1^3-2c_1\right)x^3+\cdots$$

Proof. Indeed, in this case

$$\alpha=-\frac{1}{12},\beta=\frac{1}{8}.$$

□

Corollary 55. *For no $i, j = 1, 2, ..., 10$,*

$$^{\mathcal{I}}\mathcal{F}_i = \mathcal{F}_j$$

holds.

Proof. We have

$$^{\mathcal{I}}\mathcal{A}(1, 1-x) = 1 - \frac{x}{2} - \frac{x^2}{12} - \frac{x^3}{24} + ...$$

$$^{\mathcal{I}}\mathcal{G}(1, 1-x) = 1 - \frac{x}{2} + \frac{x^2}{24} + \frac{x^3}{48} + ...$$

$$^{\mathcal{I}}\mathcal{H}(1, 1-x) = 1 - \frac{x}{2} + \frac{x^2}{6} + \frac{x^3}{12} + ...$$

$$^{\mathcal{I}}\mathcal{C}(1, 1-x) = 1 - \frac{x}{2} - \frac{x^2}{3} - \frac{x^3}{6} + ...$$

$$^{\mathcal{I}}\mathcal{F}_5(1, 1-x) = 1 - \frac{x}{2} - \frac{5x^2}{24} - \frac{x^3}{6} + ...$$

$$^{\mathcal{I}}\mathcal{F}_6(1, 1-x) = 1 - \frac{x}{2} - \frac{5x^2}{24} - \frac{x^3}{24} + ...$$

$$^{\mathcal{I}}\mathcal{F}_7(1, 1-x) = 1 - x^2 - \frac{x^3}{3} + ...$$

$$^{\mathcal{I}}\mathcal{F}_8(1, 1-x) = 1 - x^2 + \frac{2x^3}{3} + ...$$

$$^{\mathcal{I}}\mathcal{F}_9(1, 1-x) = 1 - x + x^2 - \frac{x^3}{3} + ...$$

$$^{\mathcal{I}}\mathcal{F}_{10}(1, 1-x) = 1 - x + x^2 - \frac{4x^3}{3} + ...$$

At least one of the coefficients from the left side $^{\mathcal{I}}\mathcal{F}_i$ is different from the corresponding coefficient of the mean from the right side $\mathcal{F}_j$. □

Chapter 3

Double sequences

ABSTRACT

We study general Archimedean and Gaussian double sequences, constructed by using means. We study their properties, including conditions for ensuring their convergence.

In the Introduction, we defined two types of double sequences. Beginning with two positive numbers a and b, which we also denoted by $a_o = a$ and $b_o = b$, we defined a double sequence $(a_n)_{n\geq 0}$ and $(b_n)_{n\geq 0}$, by

$$a_{n+1} = \mathcal{H}(a_n, b_n) \text{ and } b_{n+1} = \mathcal{G}(a_{n+1}, b_n)\ ,\ n \geq 0, \tag{3.1}$$

and another by

$$a_{n+1} = \mathcal{A}(a_n, b_n),\ b_{n+1} = \mathcal{G}(a_n, b_n),\ n \geq 0. \tag{3.2}$$

As we saw, the sequences $(a_n)_{n\geq 0}$ and $(b_n)_{n\geq 0}$ are monotonously convergent to a common limit which we denoted by $\mathcal{H} \boxtimes \mathcal{G}(a, b)$ in the case of the relations (3.1) and by $\mathcal{A} \otimes \mathcal{G}(a, b)$ in that of the relations (3.2). The double sequence will be called Archimedean in the first case and Gaussian in the second case.

In what follows we shall study some properties of general Archimedean and of Gaussian double sequences. The means $\mathcal{A}$, $\mathcal{G}$, and $\mathcal{H}$ will be replaced by arbitrary means M and N. Minimal conditions on these means to ensure the convergence of double sequences are determined.

3.1 ARCHIMEDEAN DOUBLE SEQUENCES

As we saw, Archimedes' polygonal method of evaluation of π was interpreted in Phillips (1981) and Phillips (2000) as a double sequence (3.1). In Schoenberg (1982) and Miel (1983) a posthumous work of R. Descartes (1596–1650) is analyzed, where the polygonal method is regarded from a new point of view. Descartes' approach consisted of doubling the number of sides of inscribed and circumscribed regular polygons while keeping the perimeter constant. As they show, this algorithm leads to the double sequence

$$a_{n+1} = \mathcal{A}(a_n, b_n) \text{ and } b_{n+1} = \mathcal{G}(a_{n+1}, b_n)\ ,\ n \geq 0, \tag{3.3}$$

which is also of Archimedean type.

Means in Mathematical Analysis. http://dx.doi.org/10.1016/B978-0-12-811080-5.00003-7

More generally, let us consider two means M and N defined on the interval J and two initial values $a, b \in J$.

Definition 18. The pair of sequences $(a_n)_{n\geq 0}$ and $(b_n)_{n\geq 0}$ defined by

$$a_{n+1} = M(a_n, b_n) \text{ and } b_{n+1} = N(a_{n+1}, b_n) \text{ , } n \geq 0, \quad (3.4)$$

where $a_0 = a, b_0 = b$, is called an **Archimedean double sequence**.

The following result was given in Costin and Toader (2004).

Lemma 15. *For every means M and N defined on the interval J and every two initial values $a, b \in J$, the sequences $(a_n)_{n\geq 0}$ and $(b_n)_{n\geq 0}$ defined by (3.4) are monotonically convergent.*

Proof. If $a \leq b$, we can show by induction that

$$a_n \leq a_{n+1} \leq b_{n+1} \leq b_n, \; n = 0, 1, \ldots \quad (3.5)$$

Indeed, assume that $a_n \leq b_n$ (which holds for $n = 0$). From (3.4) and the definition of means we have

$$a_n \leq a_{n+1} = M(a_n, b_n) \leq b_n.$$

Then, by the same reason, we have

$$a_{n+1} \leq b_{n+1} = N(a_{n+1}, b_n) \leq b_n$$

and (3.5) is proved. Therefore, $(a_n)_{n\geq 0}$ is increasing and bounded above by $b = b_0$, thus it has a limit $\alpha(a, b) \leq b$. Similarly, $(b_n)_{n\geq 0}$ is monotonically decreasing, bounded below by $a = a_0$, and has a limit $\beta(a, b) \geq a$. The case $a > b$ is similar and this completes the proof. □

Remark 68. The trivial example:

$$a_{n+1} = \Pi_1(a_n, b_n) = a_n, \; b_{n+1} = \Pi_2(a_{n+1}, b_n) = b_n, \; n \geq 0,$$

shows that, without some auxiliary assumptions on the means M and N, the sequences $(a_n)_{n\geq 0}$ and $(b_n)_{n\geq 0}$ can have different limits.

Definition 19. The mean M is **compoundable in the sense of Archimedes** (or **A-compoundable**) with the mean N if the sequences $(a_n)_{n\geq 0}$ and $(b_n)_{n\geq 0}$ defined by (3.4) are convergent to a common limit $M \boxtimes N(a, b)$ for each $a, b \in J$.

Remark 69. From the proof of the previous lemma we deduce that

$$a \wedge b \leq M \boxtimes N(a,b) \leq a \vee b \,, \; \forall a,b \in J,$$

that is, if M is A-compoundable with N, then $M \boxtimes N$ is a mean on J.

Definition 20. The mean $M \boxtimes N$ is called **Archimedean compound mean** (or **A-compound mean**) of M and N, and $\boxtimes$ is called the **Archimedean product**.

A rather general result was proved in Foster and Phillips (1984).

Theorem 71. *If the means M and N are continuous, symmetrical, and strict, then M is A-compoundable with N.*

These hypotheses were later weakened in Costin and Toader (2004) by proving that it is enough that only one of the two means have some properties.

Theorem 72. *If the mean M is continuous and strict at the left, or N is continuous and strict at the right, then M is A-compoundable with N.*

Proof. Assume that M is continuous and strict at the left. From the previous lemma we deduce that for every $a, b \in J$, the sequences $(a_n)_{n\geq 0}$ and $(b_n)_{n\geq 0}$, defined by (3.4), have the limits α respectively β. From the first relation of (3.4) and the continuity of M we deduce that $\alpha = M(\alpha, \beta)$. As M is strict at the left it follows that $\alpha = \beta$ for every $a, b \in J$. The case for N continuous and strict at the right is similar and so the proof is complete. □

Remark 70. The mean Π_1 is continuous and strict at the right, but it is not strict at the left. So, it is a "good" mean for the A-composition at the right, but it is a "bad" mean for the A-composition at the left. For a similar reason, Π_2 is good for the left A-composition, but it is bad for the right A-composition. For example, Π_2 is A-compoundable with Π_1 and

$$\Pi_2 \boxtimes \Pi_1 = \Pi_2,$$

but, as we saw, Π_1 is not A-compoundable with Π_2.

3.2 DETERMINATION OF A-COMPOUND MEANS

In studying Archimedes' algorithm from the first part of the book, we have found that

$$\mathcal{H} \boxtimes \mathcal{G}(a,b) = \begin{cases} \frac{ab}{\sqrt{a^2-b^2}} \arccos\left(\frac{b}{a}\right), & \text{for } 0 < b < a \\ \frac{ab}{\sqrt{b^2-a^2}} \operatorname{arg}\cosh\left(\frac{b}{a}\right), & \text{for } 0 < a < b \end{cases}.$$

In a letter to his teacher, in 1800 Gauss suggested the study of the algorithm (3.3). Pfaff determined the common limit

$$\mathcal{A}\boxtimes\mathcal{G}(a,b)=\begin{cases}\frac{\sqrt{b^2-a^2}}{\arccos(a/b)}, & \text{for } 0<a<b\\[2ex] \frac{\sqrt{a^2-b^2}}{\operatorname{arg\,cosh}(a/b)}, & \text{for } 0<b<a\end{cases}$$

but the result was published only in 1917 in Gauss (1876–1927). In the meantime, the same result was given in Schwab (1813) and Borchardt (1881). That's why the algorithm (3.3) is called **Pfaff–Schwab–Borchardt's algorithm**.

As it is remarked in Miel (1983), Archimedes' algorithm is essentially equivalent to Pfaff–Schwab–Borchardt's algorithm. Indeed,

$$\mathcal{H}\boxtimes\mathcal{G}\left(\frac{1}{a},\frac{1}{b}\right)=\frac{1}{\mathcal{A}\boxtimes\mathcal{G}(a,b)},\quad\forall a,b>0. \tag{3.6}$$

In Beke (1927) $\mathcal{G}\boxtimes\mathcal{A}$ was determined, while the complete table of A-compound means $M\boxtimes N$ for $M,N\in\{\mathcal{A},\mathcal{G},\mathcal{H}\}$ was given in Foster and Phillips (1984a). First of all, it is easy to prove that

$$\mathcal{A}\boxtimes\mathcal{A}=\mathcal{A}_{1/3}$$

(see the proof of a more general result at the end of this section). Then, they remarked that $\mathcal{G}$ and $\mathcal{H}$ are quasi-arithmetic means. To determine $\mathcal{A}(f)\boxtimes\mathcal{A}(f)$ for an arbitrary continuous bijection f, we have to consider the double sequence

$$f(a_{n+1})=\mathcal{A}(f(a_n),f(b_n)),\ f(b_{n+1})=\mathcal{A}(f(a_{n+1}),f(b_n)),\ \forall n\geq 0.$$

The common limit of the sequences $(f(a_n))_{n\geq 0}, (f(b_n))_{n\geq 0}$ is $\mathcal{A}_{1/3}(a_0,b_0)$, thus

$$\mathcal{A}(f)\boxtimes\mathcal{A}(f)=\mathcal{A}_{1/3}(f), \tag{3.7}$$

which gives

$$\mathcal{G}\boxtimes\mathcal{G}=\mathcal{G}_{1/3},\ \mathcal{H}\boxtimes\mathcal{H}=\mathcal{H}_{1/3}.$$

On the other hand, the method from (3.6) can be generally used and we get

$$\mathcal{H}\boxtimes\mathcal{A}\left(\frac{1}{a},\frac{1}{b}\right)=\frac{1}{\mathcal{A}\boxtimes\mathcal{H}(a,b)},\quad\forall a,b>0$$

and

$$\mathcal{G}\boxtimes\mathcal{A}\left(\frac{1}{a},\frac{1}{b}\right)=\frac{1}{\mathcal{G}\boxtimes\mathcal{H}(a,b)},\quad\forall a,b>0.$$

So, to complete the desired list, we have to determine $\mathcal{G}\boxtimes\mathcal{H}$ and $\mathcal{A}\boxtimes\mathcal{H}$.

Given the initial values a and b, we denote $a_0 = a, b_0 = b$ and construct the sequences

$$a_{n+1} = \sqrt{a_n b_n}, \quad b_{n+1} = \frac{2a_{n+1}b_n}{a_{n+1} + b_n}, \quad n \geq 0.$$

The determination of $\mathcal{G} \boxtimes \mathcal{H}(a, b)$ is analogous to the one from the Archimedean process, which has the two means transposed. As in that case, if $0 < a < b$ then we can verify that

$$a_n = 2^{n-1} \mu \sin \frac{\theta}{2^{n-1}}, \quad b_n = 2^n \mu \tan \frac{\theta}{2^n}, \quad n > 0$$

where

$$\frac{a}{b} = \cos^2 \theta, \quad \mu = \frac{b}{\sqrt{\frac{b}{a} - 1}}.$$

It follows that

$$\mathcal{G} \boxtimes \mathcal{H}(a, b) = \frac{b\sqrt{a}}{\sqrt{b - a}} \arccos \sqrt{\frac{a}{b}}.$$

If $0 < b < a$, we need to replace sin, tan, cos by the corresponding hyperbolic functions and redefine

$$\mu = \frac{b\sqrt{a}}{\sqrt{a - b}}.$$

In the final case, for the initial values a and b, we denote $a_0 = a, b_0 = b$ and construct the sequences

$$a_{n+1} = \frac{a_n + b_n}{2}, \quad b_{n+1} = \frac{2a_{n+1}b_n}{a_{n+1} + b_n}, \quad n \geq 0.$$

From these relations, we deduce that the limit $\mathcal{A} \boxtimes \mathcal{H}$ is homogeneous. Thus, we can take

$$a = 1 + x, \; b = 1, \; x > -1.$$

By induction it follows that, for any $n \geq 1$,

$$a_n = \frac{\prod_{r=1}^{n} (2^{2r-1} + x)}{2^n \prod_{r=1}^{n-1} (2^{2r} + x)}, \quad b_n = 2^n \prod_{r=1}^{n} \frac{2^{2r-1} + x}{2^{2r} + x}.$$

So the limit can be expressed by the infinite product

$$\mathcal{A} \boxtimes \mathcal{H}(1+x,1) = \prod_{r=1}^{\infty} \frac{1+\frac{2x}{4^r}}{1+\frac{x}{4^r}}.$$

This result is also used in Foster and Phillips (1984a) for the determination of the limit as an infinite series.

Using the method from (3.7), in Costin and Toader (2004) the following general result is given.

Theorem 73. *If the mean M is A-compoundable with the mean N, then for every bijection f, which has a continuous inverse, the mean $M(f)$ is A-compoundable with the mean $N(f)$ and*

$$M(f) \boxtimes N(f) = M \boxtimes N(f).$$

Proof. Given the initial values a and b, we construct the sequences $(a_n)_{n\geq 0}$ and $(b_n)_{n\geq 0}$ by

$$a_{n+1} = M(f)(a_n, b_n), \quad b_{n+1} = N(f)(a_{n+1}, b_n), \; n \geq 0,$$

where $a_0 = a, \; b_0 = b$. By the definition of quasi-M means, we have to consider the sequences $(f(a_n))_{n\geq 0}$ and $(f(b_n))_{n\geq 0}$ given by

$$f(a_{n+1}) = M(f(a_n), f(b_n)), \quad f(b_{n+1}) = N(f(a_{n+1}), f(b_n)), \; n \geq 0,$$

with the initial values $f(a), f(b)$. As M is A-compoundable with N, these sequences have the common limit $M \boxtimes N(f(a), f(b))$. As f^{-1} is continuous, the sequences $(a_n)_{n\geq 0}$ and $(b_n)_{n\geq 0}$ have the common limit $M \boxtimes N(f)(a,b)$. □

The following result was proved in Toader (1987).

Lemma 16. *For every $\lambda, \mu \in (0,1)$ we have*

$$\mathcal{A}_\lambda \boxtimes \mathcal{A}_\mu = \mathcal{A}_{\frac{\lambda\mu}{1-\lambda+\lambda\mu}}.$$

Proof. From

$$a_{n+1} = \lambda \cdot a_n + (1-\lambda) \cdot b_n, \quad b_{n+1} = \mu \cdot a_{n+1} + (1-\mu) \cdot b_n,$$

we have also

$$b_{n+1} = \lambda \cdot \mu \cdot a_n + (1 - \lambda \cdot \mu) \cdot b_n.$$

Therefore,

$$a_{n+1} - b_{n+1} = \lambda \cdot (1-\mu) \cdot (a_n - b_n)$$

which gives

$$a_n - b_n = \lambda^n \cdot (1-\mu)^n \cdot (a-b).$$

As

$$b_{n+1} = b_n + \lambda \cdot \mu \cdot (a_n - b_n),$$

we have

$$b_{n+1} = b_n + \lambda \cdot \mu \cdot \lambda^n \cdot (1-\mu)^n \cdot (a-b).$$

Adding these relations from 0 to $n-1$, we get

$$b_n = b + \lambda \cdot \mu \cdot (a-b) \cdot \frac{1-\lambda^n \cdot (1-\mu)^n}{1-\lambda \cdot (1-\mu)}.$$

So

$$\lim_{n\to\infty} b_n = b + \frac{\lambda \cdot \mu \cdot (a-b)}{1-\lambda \cdot (1-\mu)},$$

which proves the desired result. □

Corollary 56. *For every $m \in \mathbb{R}$, $\lambda, \mu \in (0,1)$ we have*

$$\mathcal{P}_{m,\lambda} \boxtimes \mathcal{P}_{m,\mu} = \mathcal{P}_{m,\frac{\lambda\mu}{1-\lambda+\lambda\mu}}.$$

Proof. Taking $f = e_m$ and using the previous result, we get the relation. □

3.3 RATE OF CONVERGENCE OF AN ARCHIMEDEAN DOUBLE SEQUENCE

In the case of classical Archimedean algorithm, we have shown that the error of the sequences $(a_n)_{n\geq 0}$ and $(b_n)_{n\geq 0}$ tends to zero asymptotically like $1/4^n$. In Foster and Phillips (1984) it is proved that this result is valid in the case of A-composition of arbitrary differentiable symmetric means. For the general case, the following evaluation is given in Costin (2004).

Let us consider two means M and N given on the interval J and two initial values $a, b \in J$. Define the pair of sequences $(a_n)_{n\geq 0}$ and $(b_n)_{n\geq 0}$ by

$$a_{n+1} = M(a_n, b_n) \text{ and } b_{n+1} = N(a_{n+1}, b_n),\ n \geq 0,$$

where $a_0 = a, b_0 = b$. Denote

$$\alpha = M \boxtimes N(a,b).$$

Theorem 74. *If the means M and N have continuous partial derivatives up to the second order, then the errors of the sequences $(a_n)_{n\geq 0}$ and $(b_n)_{n\geq 0}$ tend to zero asymptotically like*

$$[M_a(\alpha,\alpha)\cdot(1-N_a(\alpha,\alpha))]^n .$$

Proof. If we write

$$a_n=\alpha+\delta_n,\ b_n=\alpha+\varepsilon_n,$$

we deduce that, as $n\to\infty$,

$$\begin{aligned}\alpha+\delta_{n+1}&=M(\alpha+\delta_n,\alpha+\varepsilon_n)\\&=M(\alpha,\alpha)+M_a(\alpha,\alpha)\delta_n+M_b(\alpha,\alpha)\varepsilon_n+O(\delta_n^2+\varepsilon_n^2).\end{aligned}$$

From (2.26) we get

$$\delta_{n+1}=M_a(\alpha,\alpha)\delta_n+[1-M_a(\alpha,\alpha)]\,\varepsilon_n+O(\delta_n^2+\varepsilon_n^2). \tag{3.8}$$

Then

$$\begin{aligned}\alpha+\varepsilon_{n+1}&=N(\alpha+\delta_{n+1},\alpha+\varepsilon_n)\\&=N(\alpha,\alpha)+N_a(\alpha,\alpha)\delta_{n+1}+N_b(\alpha,\alpha)\varepsilon_n+O(\delta_{n+1}^2+\varepsilon_n^2).\end{aligned}$$

Using again (2.26) and (3.8) we have

$$\begin{aligned}\varepsilon_{n+1}=N_a(\alpha,\alpha)\,&[M_a(\alpha,\alpha)\delta_n+(1-M_a(\alpha,\alpha))\,\varepsilon_n]\\&+[1-N_a(\alpha,\alpha)]\,\varepsilon_n+O(\delta_n^2+\varepsilon_n^2),\end{aligned}$$

thus

$$\varepsilon_{n+1}=M_a(\alpha,\alpha)N_a(\alpha,\alpha)\delta_n+[1-M_a(\alpha,\alpha)N_a(\alpha,\alpha)]\,\varepsilon_n+O(\delta_n^2+\varepsilon_n^2). \tag{3.9}$$

Subtracting (3.9) from (3.8) we get

$$\delta_{n+1}-\varepsilon_{n+1}=M_a(\alpha,\alpha)\,[1-N_a(\alpha,\alpha)]\,(\delta_n-\varepsilon_n)+O(\delta_n^2+\varepsilon_n^2).$$

On the other hand, from the monotonicity of $(a_n)_{n\geq 0}$ and $(b_n)_{n\geq 0}$ we can assume that $\delta_n>0$ and $\varepsilon_n<0$ for all $n>0$. The cases when $\delta_n<0$ and $\varepsilon_n>0$ can be treated similarly. We have

$$\frac{\varepsilon_n-\varepsilon_{n+1}}{\delta_n-\delta_{n+1}}=\frac{M_a(\alpha,\alpha)N_a(\alpha,\alpha)\,(\varepsilon_n-\delta_n)+O(\delta_n^2+\varepsilon_n^2)}{[1-M_a(\alpha,\alpha)]\,(\delta_n-\varepsilon_n)+O(\delta_n^2+\varepsilon_n^2)},$$

thus

$$\varepsilon_n - \varepsilon_{n+1} = \frac{M_a(\alpha,\alpha)N_a(\alpha,\alpha)}{M_a(\alpha,\alpha)-1}(\delta_n - \delta_{n+1}) + (\delta_n - \delta_{n+1})O(|\delta_n| + |\varepsilon_n|).$$

Replacing n by $n+1, n+2, ..., n+p-1$ ($p \in \mathbb{N}$), adding and using the fact that δ_n and ε_n tend monotonically to zero, we obtain

$$\varepsilon_n - \varepsilon_{n+p} = \frac{M_a(\alpha,\alpha)N_a(\alpha,\alpha)}{M_a(\alpha,\alpha)-1}\left(\delta_n - \delta_{n+p}\right) + (\delta_n - \delta_{n+p})O(|\delta_n| + |\varepsilon_n|).$$

Letting $p \to \infty$ we obtain

$$\varepsilon_n = \frac{M_a(\alpha,\alpha)N_a(\alpha,\alpha)}{M_a(\alpha,\alpha)-1}\delta_n + O(\delta_n^2 + \varepsilon_n^2).$$

Using (3.8) we deduce that

$$\delta_{n+1} = M_a(\alpha,\alpha)\left[1 - N_a(\alpha,\alpha)\right]\delta_n + O(\delta_n^2)$$

and from (3.9) we have

$$\varepsilon_{n+1} = M_a(\alpha,\alpha)\left[1 - N_a(\alpha,\alpha)\right]\varepsilon_n + O(\varepsilon_n^2).$$ □

Remark 71. Some evaluations and conclusions related to the previous theorem may also be found in Foster and Phillips (1985). In the case of symmetric means, the result was proved in Foster and Phillips (1984). In this case, we saw in (2.27) that

$$M_a(\alpha,\alpha) = N_a(\alpha,\alpha) = \frac{1}{2}, \ \forall \alpha \in J$$

and we get the following.

Corollary 57. *If the means M and N are symmetric and have continuous partial derivatives up to the second order, then the error of the sequences $(a_n)_{n\geq 0}$ and $(b_n)_{n\geq 0}$ tends to zero asymptotically like $1/4^n$.*

3.4 ACCELERATION OF THE CONVERGENCE

In the case of classical Archimedean algorithm it can be proved, as it was shown in Foster and Phillips (1983) using MacLaurin's series, that if we define

$$c_n = \frac{a_n + 2b_n}{3},$$

then $(c_n)_{n\geq 0}$ converges faster than $(a_n)_{n\geq 0}$ or $(b_n)_{n\geq 0}$. Indeed, we have

$$a_n = 2^n\lambda \cdot \tan\frac{\theta}{2^n} \text{ and } b_n = 2^n\lambda \cdot \sin\frac{\theta}{2^n}, \ \forall n > 0.$$

Then

$$a_n = 2^n\lambda \cdot \left(\frac{\theta}{2^n} + \frac{\theta^3}{3 \cdot 2^{3n}} + \ldots\right) = \lambda\theta + \lambda\frac{\theta^3}{3 \cdot 4^n} + \ldots$$

and

$$b_n = 2^n\lambda \cdot \left(\frac{\theta}{2^n} - \frac{\theta^3}{6 \cdot 2^{3n}} + \ldots\right) = \lambda\theta - \lambda\frac{\theta^3}{6 \cdot 4^n} + \ldots$$

but

$$c_n = \frac{2^n\lambda}{3} \cdot \left(3\frac{\theta}{2^n} + \frac{3\theta^5}{20 \cdot 2^{5n}} + \ldots\right) = \lambda\theta + \lambda\frac{\theta^5}{20 \cdot 16^n} + \ldots$$

It is remarkable that in Snell (1621), the sequence $(c_n)_{n\geq 0}$ was used to compute π to 34 decimal places. Of course, the author did not have access to the series for the sine and tangent. In the general case, in Foster and Phillips (1983) it was shown that the sequence $(c_n)_{n\geq 0}$ has the same property for sufficiently smooth symmetric means.

For non-symmetric means, the way of constructing a sequence $(c_n)_{n\geq 0}$ with the above property is indicated in Costin (2004a).

Theorem 75. *If the means M and N have continuous partial derivatives up to the third order and verify the relation*

$$M_a(\alpha,\alpha)\left[M_a(\alpha,\alpha) - 1\right]\left[N_{aa}(\alpha,\alpha) - N_{bb}(\alpha,\alpha)\right] = N_a(\alpha,\alpha)\left[M_{aa}(\alpha,\alpha) - M_{bb}(\alpha,\alpha)\right], \tag{3.10}$$

then the sequence $(c_n)_{n\geq 0}$ given by

$$c_n = \frac{M_a(\alpha,\alpha)N_a(\alpha,\alpha)a_n + \left[1 - M_a(\alpha,\alpha)\right]b_n}{1 - M_a(\alpha,\alpha) + M_a(\alpha,\alpha)N_a(\alpha,\alpha)}.$$

converges faster than $(a_n)_{n\geq 0}$ and $(b_n)_{n\geq 0}$.

Proof. Keeping the notation from the previous paragraph, we have this time

$$\alpha + \delta_{n+1} = M(\alpha + \delta_n, \alpha + \varepsilon_n) = M(\alpha,\alpha) + M_a(\alpha,\alpha)\delta_n + M_b(\alpha,\alpha)\varepsilon_n + \frac{1}{2}\left[M_{aa}(\alpha,\alpha)\delta_n^2 + 2M_{ab}(\alpha,\alpha)\delta_n\varepsilon_n + M_{bb}(\alpha,\alpha)\varepsilon_n^2\right] + O(|\delta_n|^3 + |\varepsilon_n|^3).$$

Using (2.28) we get

$$\delta_{n+1} = M_a(\alpha,\alpha)\delta_n + \left[1 - M_a(\alpha,\alpha)\right]\varepsilon_n + \frac{1}{2}\left[M_{aa}(\alpha,\alpha)\delta_n - M_{bb}(\alpha,\alpha)\varepsilon_n\right](\delta_n - \varepsilon_n) + O(|\delta_n|^3 + |\varepsilon_n|^3), \tag{3.11}$$

thus

$$\delta_{n+1} - \delta_n = (\varepsilon_n - \delta_n)\left[1 - M_a(\alpha,\alpha) - \frac{1}{2}\left(M_{aa}(\alpha,\alpha)\delta_n - M_{bb}(\alpha,\alpha)\varepsilon_n\right)\right] + O(|\delta_n|^3 + |\varepsilon_n|^3).$$

Similarly we get

$$\varepsilon_{n+1} - \varepsilon_n = (\delta_{n+1} - \varepsilon_n)\left[N_a(\alpha,\alpha) + \frac{1}{2}\left(N_{aa}(\alpha,\alpha)\delta_{n+1} - N_{bb}(\alpha,\alpha)\varepsilon_n\right)\right] + O(|\delta_n|^3 + |\varepsilon_n|^3).$$

Taking into account that

$$\delta_{n+1} - \varepsilon_n = (\delta_n - \varepsilon_n)\left[M_a(\alpha,\alpha) + \frac{1}{2}\left(M_{aa}(\alpha,\alpha)\delta_n - M_{bb}(\alpha,\alpha)\varepsilon_n\right)\right] + O(|\delta_n|^3 + |\varepsilon_n|^3),$$

we have also

$$\begin{aligned}\varepsilon_{n+1} - \varepsilon_n = (\delta_n - \varepsilon_n)\Big[&M_a(\alpha,\alpha)N_a(\alpha,\alpha) \\ &+\frac{N_a(\alpha,\alpha)}{2}(M_{aa}(\alpha,\alpha)\delta_n - M_{bb}(\alpha,\alpha)\varepsilon_n) + \frac{M_a(\alpha,\alpha)}{2}(N_{aa}(\alpha,\alpha) - N_{bb}(\alpha,\alpha)) \\ &+\frac{1}{2}M_a^2(\alpha,\alpha)N_{aa}(\alpha,\alpha)(\delta_n - \varepsilon_n)\Big] + O(|\delta_n|^3 + |\varepsilon_n|^3).\end{aligned}$$

For some real λ let us denote

$$\Delta_n = \lambda\delta_n + (1-\lambda)\varepsilon_n.$$

We have

$$\begin{aligned}\Delta_{n+1} - \Delta_n = (\varepsilon_n - \delta_n)\Big\{&\lambda\left[1 - M_a(\alpha,\alpha) - \frac{1}{2}(M_{aa}(\alpha,\alpha)\delta_n - M_{bb}(\alpha,\alpha)\varepsilon_n)\right] \\ &+(\lambda-1)\Big[M_a(\alpha,\alpha)N_a(\alpha,\alpha) + \frac{N_a(\alpha,\alpha)}{2}(M_{aa}(\alpha,\alpha)\delta_n - M_{bb}(\alpha,\alpha)\varepsilon_n) \\ &+\frac{M_a(\alpha,\alpha)}{2}\cdot\varepsilon_n(N_{aa}(\alpha,\alpha) - N_{bb}(\alpha,\alpha)) + \frac{1}{2}M_a^2(\alpha,\alpha)N_{aa}(\alpha,\alpha)(\delta_n - \varepsilon_n)\Big]\Big\} \\ &+O(|\delta_n|^3 + |\varepsilon_n|^3).\end{aligned}$$

We choose λ such that

$$\lambda[1 - M_a(\alpha,\alpha)] + (\lambda - 1)M_a(\alpha,\alpha)N_a(\alpha,\alpha) = 0,$$

thus

$$\lambda = \frac{M_a(\alpha,\alpha)N_a(\alpha,\alpha)}{1 - M_a(\alpha,\alpha) + M_a(\alpha,\alpha)N_a(\alpha,\alpha)}.$$

We get

$$\Delta_{n+1} - \Delta_n = \frac{\varepsilon_n - \delta_n}{2[1 - M_a(\alpha,\alpha) + M_a(\alpha,\alpha)N_a(\alpha,\alpha)]} \{\varepsilon_n[N_a(\alpha,\alpha)M_{bb}(\alpha,\alpha)$$
$$+ M_a(\alpha,\alpha)(M_a(\alpha,\alpha) - 1)(N_{aa}(\alpha,\alpha) - N_{bb}(\alpha,\alpha))] - \delta_n N_a(\alpha,\alpha)M_{aa}(\alpha,\alpha)$$
$$+ (M_a(\alpha,\alpha) - 1)\, M_a^2(\alpha,\alpha)N_{aa}(\alpha,\alpha)\,(\delta_n - \varepsilon_n)\Big\} + O(|\delta_n|^3 + |\varepsilon_n|^3).$$

The condition (3.10) gives

$$\Delta_{n+1} - \Delta_n = \frac{(\varepsilon_n - \delta_n)^2}{2[1 - M_a(\alpha,\alpha) + M_a(\alpha,\alpha)N_a(\alpha,\alpha)]} [N_a(\alpha,\alpha)M_{aa}(\alpha,\alpha)$$
$$- (M_a(\alpha,\alpha) - 1)\, M_a^2(\alpha,\alpha)N_{aa}(\alpha,\alpha)\Big] + O(|\delta_n|^3 + |\varepsilon_n|^3),$$

thus the sequence $(c_n)_{n\geq 0}$ converges faster than $(a_n)_{n\geq 0}$ or $(b_n)_{n\geq 0}$. □

Remark 72. The condition (3.10) is satisfied by symmetric means because for them

$$M_{aa}(\alpha,\alpha) = M_{bb}(\alpha,\alpha), \ \forall \alpha \in J.$$

Also, in this case

$$M_a(\alpha,\alpha) = \frac{1}{2}, \ \forall \alpha \in J,$$

so that we have the following.

Corollary 58. *If the symmetric means M and N have continuous partial derivatives up to the third order then the sequence* $(c_n)_{n\geq 0}$ *given by*

$$c_n = \frac{a_n + 2b_n}{3}$$

converges faster than $(a_n)_{n\geq 0}$ *and* $(b_n)_{n\geq 0}$, *namely the sequence of its errors verifies the relation*

$$\Delta_{n+1} - \Delta_n = \frac{(\varepsilon_n - \delta_n)^2}{12} [4M_{aa}(\alpha,\alpha) + N_{aa}(\alpha,\alpha)] + O(|\delta_n|^3 + |\varepsilon_n|^3).$$

Remark 73. There are also non-symmetric means with the property

$$M_{aa}(\alpha,\alpha) = M_{bb}(\alpha,\alpha), \ \forall \alpha \in J.$$

For them, the condition (3.10) is also satisfied. For instance, we have

$$\left(\mathcal{P}_{m,\lambda}\right)_a(\alpha,\alpha)=\lambda$$

and

$$\left(\mathcal{P}_{m,\lambda}\right)_{aa}(\alpha,\alpha)=\left(\mathcal{P}_{m,\lambda}\right)_{bb}(\alpha,\alpha)=\frac{\lambda(1-\lambda)(m-1)}{\alpha}.$$

So, if we study $\mathcal{P}_{m,\lambda}\boxtimes\mathcal{P}_{m,\mu}$, we have

$$c_n=\frac{\lambda\mu a_n+(1-\lambda)b_n}{1-\lambda+\lambda\mu}$$

and

$$\Delta_{n+1}-\Delta_n=\frac{(\varepsilon_n-\delta_n)^2\lambda\mu(1-\lambda)(m-1)(1+\lambda-\lambda\mu)}{2(1-\lambda+\lambda\mu)\alpha}+O(|\delta_n|^3+|\varepsilon_n|^3).$$

3.5 GAUSSIAN DOUBLE SEQUENCES

The arithmetic–geometric process of Gauss was also generalized for arbitrary means. Considering two means M and N defined on the interval J and two initial values $a,b\in J$, we have the following.

Definition 21. The pair of sequences $(a_n)_{n\geq 0}$ and $(b_n)_{n\geq 0}$ defined by

$$a_{n+1}=M(a_n,b_n) \text{ and } b_{n+1}=N(a_n,b_n),\ n\geq 0, \tag{3.12}$$

where $a_0=a,\ b_0=b$, is called a **Gaussian double sequence**.

Remark 74. An Archimedean double sequence can be considered also as a Gaussian double sequence defined by

$$a_{n+1}=M(a_n,b_n) \text{ and } b_{n+1}=N(M,\Pi_2)(a_n,b_n),\ n\geq 0.$$

Remark 75. The sequences $(a_n)_{n\geq 0}$ and $(b_n)_{n\geq 0}$ defined by (3.12) are not necessarily convergent.

Example 8. Take $M=\Pi_2$ and $N=\Pi_1$. It follows that $a_{n+1}=b_n$ and $b_{n+1}=a_n,\ n\geq 0$, thus the sequences $(a_n)_{n\geq 0}$ and $(b_n)_{n\geq 0}$ are divergent unless $a=b$.

However, the following property was proved in Costin and Toader (2006).

Lemma 17. *Given a Gaussian double sequence (3.12), if we denote*

$$\underline{a_n}=a_n\wedge b_n \textit{ and } \overline{b_n}=a_n\vee b_n,\ n\geq 0,$$

then the sequences $\left(\underline{a_n}\right)_{n\geq 0}$ *and* $\left(\overline{b_n}\right)_{n\geq 0}$ *are monotonously convergent.*

Proof. For each $n \geq 0$, we have

$$\underline{a_n} \leq a_{n+1} = M(a_n, b_n) \leq \overline{b_n}$$

and

$$\underline{a_n} \leq b_{n+1} = N(a_n, b_n) \leq \overline{b_n},$$

thus

$$\underline{a_n} \leq \underline{a_{n+1}} \leq \overline{b_{n+1}} \leq \overline{b_n}$$

and the conclusion follows as in the corresponding lemma for the Archimedean case. □

Remark 76. If the sequences $\left(\underline{a_n}\right)_{n\geq 0}$ and $\left(\overline{b_n}\right)_{n\geq 0}$ are convergent to a common limit, then the sequences $(a_n)_{n\geq 0}$ and $(b_n)_{n\geq 0}$ are also convergent to the same limit which lies between $a \wedge b$ and $a \vee b$.

Definition 22. The mean M is **compoundable in the sense of Gauss** (or **G-compoundable**) with the mean N if the sequences $(a_n)_{n\geq 0}$ and $(b_n)_{n\geq 0}$ defined by (3.12) are convergent to a common limit, denoted $M \otimes N(a, b)$, for each $a, b \in J$. If M is G-compoundable with N and also N is G-compoundable with M we say that M and N are **G**-compoundable.

Following the previous remark, the function $M \otimes N$ defines a mean which is called **Gaussian compound mean** (or **G-compound mean**) and $\otimes$ is called **Gaussian product**.

In the case of G-composition the study of convergence is more complicated. The first results were given for homogeneous means in Andreoli (1957a), Myrberg (1958), Allasia (1969–1970), and Tricomi (1975). Other results were published for the case of comparable, strict, and continuous means in Schoenberg (1982), Foster and Phillips (1985), Wimp (1985), Borwein and Borwein (1987a), and Toader (1987). The results and the proofs are very similar with those of the original Gauss' algorithm.

Minimal conditions on the means M and N were given in Costin and Toader (2006).

Theorem 76. *If the means M and N are comparable, M is continuous and strict at the left, or N is continuous and strict at the right, then they are G-compoundable.*

Proof. If $M \leq N$ and $a < b$, we have

$$a \leq a_1 = M(a, b) \leq N(a, b) = b_1 \leq b.$$

By induction, assuming that $a_n \le b_n$, it follows that

$$a_n \le a_{n+1} = M(a_n, b_n) \le N(a_n, b_n) = b_{n+1} \le b_n.$$

Thus $(a_n)_{n\ge 0}$ increases while $(b_n)_{n\ge 0}$ decreases. Since each of these sequences is bounded by a and b, both converge, say, to α and β, respectively. If M is assumed continuous then $\alpha = M(\alpha, \beta)$. But M is also strict at the left, so that $\alpha = \beta$. Similarly, if N is continuous and strict at the right, we have $\beta = N(\alpha, \beta)$, thus again $\alpha = \beta$. If $a > b$, we change only the first inequalities

$$b \le a_1 = M(a, b) \le N(a, b) = b_1 \le a.$$

If $M \ge N$, we have to exchange the roles of a_n and b_n. In all the cases $M \otimes N$ exists and satisfies

$$M \wedge N \le M \otimes N \le M \vee N. \qquad \square$$

Theorem 77. *If the means M and N are in the relation $M \prec N$, M is continuous and strict at the left, or N is continuous and strict at the right, then M is G-compoundable with N.*

Proof. If $a < b$ we have $M(a, b) \le N(a, b)$ and the proof may be continued as that of the previous theorem. If $a > b$, we have $M(a, b) \ge N(a, b)$ and we interchange the roles of a_n and b_n. □

Theorem 78. *If the means M and N are in the relation $N \prec M$, M is continuous and strict at the right or N is continuous and strict at the left, then M is G-compoundable with N.*

Proof. This time, if $a < b$ we have $M(a, b) \ge N(a, b)$, so that

$$b_0 = b \ge a_1 = M(a, b) \ge N(a, b) = b_1 \ge a = a_0,$$

or generally, by induction,

$$b_0 \ge a_1 \ge b_2 \ge a_3 \ge \dots \ge a_{2n-1} \ge b_{2n} \ge \dots \ge a_{2n} \ge b_{2n-1} \ge \dots \ge b_1 \ge a_0.$$

Hence the sequences $(a_{2n+1})_{n\ge 0}$ and $(b_{2n})_{n\ge 0}$ have the same limit, say α, while the sequences $(a_{2n})_{n\ge 0}$ and $(b_{2n+1})_{n\ge 0}$ have the same limit, say β. If M is continuous and strict at the right, from

$$a_{2n+1} = M(a_{2n}, b_{2n})$$

we get $\alpha = M(\beta, \alpha)$, thus $\alpha = \beta$. If N is continuous and strict at the left, we use the relation

$$b_{2n+1} = N(a_{2n}, b_{2n})$$

to get the same conclusion $\alpha = \beta$. The case $a > b$ can be treated similarly. □

In Toader (1990) the comparability for symmetric means was renounced. The proof is very simple and it is based on the use of the means $M \wedge N$ and $M \vee N$. With a more sophisticated method the result was proved for non-symmetric means in Foster and Phillips (1986).

Theorem 79. *If the means M and N are continuous and strict at the left, then M and N are G-compoundable.*

Remark 77. We have also a variant for means which are strict at the right. As the example of the means Π_1 and Π_2 which are not G-compoundable (in any order) shows, the result is not valid if we assume one mean to be strict at the left and the other strict at the right. But, as it was proved in Toader (1990a) (but also in Daróczy and Páles, 2002 and Matkowski, 2006), we can G-compose a strict mean with any mean. The proof is very similar to that of Foster and Phillips (1986).

Theorem 80. *If one of the means M and N is continuous and strict, then M and N are G-compoundable.*

Proof. From (3.12) it follows that the sequences $(a_n)_{n\geq 0}$ and $(b_n)_{n\geq 0}$ lie in the closed interval determined by a and b. By the Bolzano–Weierstrass theorem, there are the subsequences $\left(a_{n_k}\right)_{k\geq 0}$, $\left(b_{n_k}\right)_{k\geq 0}$ and the points $\alpha, \beta, \alpha', \beta'$ such that

$$\lim_{k\to\infty} a_{n_k} = \alpha,\ \lim_{k\to\infty} b_{n_k} = \beta,\ \lim_{k\to\infty} a_{n_{k+1}} = \alpha',\ \lim_{k\to\infty} b_{n_{k+1}} = \beta'.$$

We can prove that $\alpha = \beta$. Indeed, suppose $\alpha < \beta$. From (3.12) it follows that

$$\alpha \leq \alpha' \leq \beta,\ \alpha \leq \beta' \leq \beta.$$

We show that

$$\alpha' = \alpha \text{ or } \alpha' = \beta. \tag{3.13}$$

If $\alpha < \alpha' \leq \beta' \leq \beta$, we choose $0 < r < (\alpha' - \alpha)/2$ and K_r such that

$$\left|a_{n_{k+1}} - \alpha'\right| < r,\ \left|b_{n_{k+1}} - \alpha'\right| < r,\ \forall k \geq K_r.$$

Therefore,

$$a_{n_{k+1}} > \alpha' - r > (\alpha' + \alpha)/2$$

and

$$b_{n_{k+1}} > \beta' - r > (\alpha' + \alpha)/2.$$

It follows that

$$a_{n_k} > (\alpha' + \alpha)/2 > \alpha, \ \forall k \geq K_r,$$

which is inconsistent with the hypothesis that $\lim_{k\to\infty} a_{n_k} = \alpha$. If $\alpha \leq \beta' \leq \alpha' < \beta$, we obtain a similar contradiction by choosing $0 < r < (\beta - \alpha')/2$. Analogously we can prove that

$$\beta' = \alpha \text{ or } \beta' = \beta \tag{3.14}$$

holds. Now, if M is continuous and strict, using (3.13) we get

$$\alpha = M(\alpha, \beta) \text{ or } \beta = M(\alpha, \beta),$$

thus $\alpha = \beta$. If N is continuous and strict, we use (3.14) to arrive at the same conclusion. The hypothesis $\alpha > \beta$ gives analogously $\alpha = \beta$. Hence

$$\lim_{k\to\infty} a_{n_k} = \lim_{k\to\infty} b_{n_k} = \alpha,$$

which leads to

$$\lim_{n\to\infty} a_n = \lim_{n\to\infty} b_n = \alpha.$$

□

Remark 78. Using this result, one can G-compose a "good" mean (that is, a continuous and strict mean) even with a "bad" one. But products like $\Pi_1 \otimes \Pi_2$, $\Pi_2 \otimes \Pi_1$, $\vee \otimes \wedge$ or $\wedge \otimes \vee$ do not exist.

Example 9. In Lehmer (1971) it is remarked that $\mathcal{A} \otimes M$ exists for every mean M and

$$\mathcal{A} \otimes M(a, b) = a_o + \frac{1}{2}\sum_{n=0}^{\infty}(b_n - a_n).$$

Indeed, $a_{k+1} - a_k = (b_k - a_k)/2$, thus

$$a_{n+1} = a_o + \sum_{k=0}^{n}(a_{k+1} - a_k) = a_o + \frac{1}{2}\sum_{k=0}^{n}(b_k - a_k).$$

A result using other type of conditions was given in Matkowski (2009).

Theorem 81. *If the means M, N are continuous and*

$$\max(M(x, y), N(x, y) - \min(M(x, y), N(x, y)) < V(x, y) - \Lambda(x, y), \ x \neq y,$$

then M and N are G-compoundable.

Remark 79. In some papers, the common limit of a double sequence is represented by an infinite product. We can mention, for example, the papers Andreoli (1957), Buzano (1965–1966), Allasia (1969–1970, 1971–1972, 1972), Allasia and Bonardo (1980), Gatteschi (1969–1970, 1982), Tricomi (1965, 1965a, 1975), or Borwein (1991).

The following properties were proved in Borwein and Borwein (1987).

Proposition 8. *If the mean M is G-compoundable with N, then the mean $M \otimes N$ is strict, homogeneous, symmetric, isotone, or continuous if each of M and N is.*

Proof. We shall use in what follows only the isotony of a G-compound mean, thus let us prove it. Assume that M and N are isotone and denote $M \otimes N = P$. For $a \leq a'$ and $b \leq b'$, define

$$a_0 = a, b_0 = b, a_{n+1} = M(a_n, b_n) \text{ and } b_{n+1} = N(a_n, b_n) \text{ for } n \geq 0$$

and

$$a'_0 = a', b'_0 = b', a'_{n+1} = M(a'_n, b'_n) \text{ and } b'_{n+1} = N(a'_n, b'_n) \text{ for } n \geq 0.$$

Using (2.3), we can prove by mathematical induction that

$$a_n \leq a'_n, \ b_n \leq b'_n \text{ for } n \geq 0.$$

Passing to the limit, we get

$$P(a, b) \leq P(a', b'),$$

which proves the conclusion. □

To establish the integral representation of $\mathcal{AGM}$, the Gauss' functional equation was used. The method was generalized in Borwein and Borwein (1987) where the following result was proved.

Theorem 82 (Invariance Principle). *Suppose that $M \otimes N$ exists and is continuous. Then $M \otimes N$ is the unique mean P which is (M, N)-invariant.*

Proof. Assuming that

$$P(M(a, b), N(a, b)) = P(a, b) \tag{3.15}$$

for all $a, b \in J$, by iteration of (3.15) it follows that

$$P(a, b) = P(a_n, b_n) = \lim_{n \to \infty} P(a_n, b_n).$$

Thus

$$P(a,b) = P(M \otimes N(a,b), M \otimes N(a,b)) = M \otimes N(a,b). \qquad \square$$

Remark 80. As in (1.26), the relation (3.15) is called **Gauss' functional equation**.

Remark 81. Writing (3.4) as

$$a_{n+1} = M(a_n, b_n), \quad b_{n+1} = N(M, \Pi_2)(a_n, b_n),$$

we have

$$M \boxtimes N = M \otimes N(M, \Pi_2).$$

So, as $M \otimes N = P$ gives $P(M, N) = P$, then $M \boxtimes N = R$ gives

$$R(M, N(M, \Pi_2)) = R. \tag{3.16}$$

Remark 82. Another construction instead that of double sequences was used in Matkowski (1999a). A mapping $\mathfrak{M} : J^2 \to J^2$ is called a **mean-type mapping** if there are two means $M, N : J^2 \to J$ such that

$$\mathfrak{M}(x, y) = (M(x, y), N(x, y)), \ \forall (x, y) \in J^2,$$

shortly $\mathfrak{M} = (M, N)$. The composition of two mean-type mappings is a mean-type mapping. So, if $\mathfrak{M}$ is a mean-type mapping then there exists the sequence of iterates of $\mathfrak{M}$. It is proved that if M and N are strict continuous means, then the sequence of iterates of $\mathfrak{M} = (M, N)$ converges to a mean-type mapping $\mathfrak{P} = (P, P)$, where P is the unique continuous (M, N)-invariant mean. A special case is that of **power mean-type mapping** $\mathfrak{P}^{[p,q]} = (P_p, P_q)$, $p, q \in \mathbb{R}$. In Kahlig and Matkowski (1997) it is proved that the composition of two power mean-type mappings is a power mean-type mapping only in the following special cases

$$\mathfrak{P}^{[p,q]} \circ \mathfrak{P}^{[r,r]} = \mathfrak{P}^{[r,r]}, \ \mathfrak{P}^{[p,p]} \circ \mathfrak{P}^{[p,0]} = \mathfrak{P}^{[p/2,p/2]},$$
$$\mathfrak{P}^{[0,0]} \circ \mathfrak{P}^{[-p,p]} = \mathfrak{P}^{[0,0]}, \ p, q, r \in \mathbb{R}.$$

3.6 DETERMINATION OF G-COMPOUND MEANS

The arithmetic–geometric G-compound mean can be represented by

$$\mathcal{A} \otimes \mathcal{G}(a,b) = \frac{\pi}{2} \cdot \left[\int_0^{\pi/2} \frac{d\theta}{\sqrt{a^2 \cos^2 \theta + b^2 \sin^2 \theta}} \right]^{-1},$$

as we saw in the first chapter. The proof is based on the Gauss's functional equation (1.18).

Only a few G-compound means can be determined by direct computation. For example, for every mean M we have

$$M \otimes M = M,$$

but also

$$\Pi_1 \otimes M = \Pi_1$$

and

$$M \otimes \Pi_2 = \Pi_2.$$

As it is shown in Borwein and Borwein (1987),

$$M \otimes \vee = \vee \otimes M = \vee \text{ and } M \otimes \wedge = \wedge \otimes M = \wedge,$$

for each strict, continuous mean M. Indeed, in each case one of the resulting sequences is constant (a or b) and the other tends monotonously to the same value.

Of course

$$N \otimes M(a,b) = M \otimes N(b,a),$$

so that, if M and N are symmetric,

$$N \otimes M = M \otimes N$$

holds. But, we have for example

$$\Pi_1 \otimes \mathcal{G} \neq \mathcal{G} \otimes \Pi_1.$$

Indeed, as we saw $\Pi_1 \otimes \mathcal{G} = \Pi_1$ but $\mathcal{G} \otimes \Pi_1 = \mathcal{G}_{2/3}$.

Looking after methods of determination of G-compound means, we can underline two general results. The first result was given in Borwein and Borwein (1987) for f a homeomorphism, and it was proved in Costin and Toader (2006) in a slightly more general form.

Theorem 83. *If the mean M is G-compoundable with the mean N, then for every bijection f, with continuous inverse, the mean $M(f)$ is G-compoundable with the mean $N(f)$ and*

$$M(f) \otimes N(f) = M \otimes N(f).$$

Let us give some applications. For every $\lambda, \mu \in (0, 1)$, we have

$$\mathcal{A}_\lambda \otimes \mathcal{A}_\mu = \mathcal{A}_{\frac{\mu}{1-\lambda+\mu}},$$

as it was proved in Toader (1987). Taking $f = e_m$ and using the previous result, we obtain the following consequence.

Corollary 59. *For every $m \in \mathbb{R}$ and $\lambda, \mu \in (0, 1)$, we have*

$$\mathcal{P}_{m,\lambda} \otimes \mathcal{P}_{m,\mu} = \mathcal{P}_{m, \frac{\mu}{1-\lambda+\mu}}.$$

For every $\lambda \in (0, 1)$ we also have

$$\mathcal{A}_\lambda \otimes \mathcal{H}_{1-\lambda} = \mathcal{G},$$

as it was proved in Toader (1991). Taking $f = e_m$ in the previous theorem we obtain another consequence.

Corollary 60. *For every $m \in \mathbb{R}$ and $\lambda \in (0, 1)$, we have*

$$\mathcal{P}_{m,\lambda} \otimes \mathcal{P}_{-m,1-\lambda} = \mathcal{G}.$$

The special case, given by $m = 1$ and $\lambda = 1/2$, can be found in Borwein and Borwein (1987).

The second method of determination of G-compound means is given by the invariance principle. For example, we have the following results.

Theorem 84. *We have*

$$M \otimes N = \mathcal{A}_\lambda(f)$$

if and only if

$$N = {}^{\mathcal{A}_\lambda(f)}M = f^{-1}\left[\frac{f(\mathcal{A}_\lambda(f)) - \lambda f(M)}{1-\lambda}\right].$$

Corollary 61. *We have*

$$M \otimes N = \mathcal{G}_\lambda$$

if and only if

$$N = {}^{\mathcal{G}(\lambda)}M = \left(\frac{\mathcal{G}_\lambda}{M^\lambda}\right)^{\frac{1}{1-\lambda}}.$$

Corollary 62. *For $m \neq 0$,*

$$M \otimes N = \mathcal{P}_{m;\lambda}$$

if and only if

$$N = {}^{\mathcal{P}(m,\lambda)}M = \left[\frac{\left(\mathcal{P}_{m;\lambda}\right)^m - \lambda M^m}{1-\lambda}\right]^{\frac{1}{m}}.$$

Remark 83. If we take $m = 0$, $p = 1/3$, and $M = \mathcal{A}_{1/3}$, or $p = 1/2, m \neq 0$, we get some exercises from Borwein and Borwein (1987).

Ninety G-compound means based on Greek means were given in Toader and Toader (2005) using also the invariance principle. Among them we have the well-known products:

$$\mathcal{A} \otimes \mathcal{H} = \mathcal{G} \text{ and } \mathcal{H} \otimes \mathcal{C} = \mathcal{A},$$

but also the new products:

$$\mathcal{F}_7 \otimes \mathcal{F}_9 = \mathcal{A} \text{ and } \mathcal{F}_8 \otimes \mathcal{F}_9 = \mathcal{G}.$$

There is no other case, non-equivalent to one of these four examples, in which all the terms be Greek means.

The following results can also be obtained using the invariance principle and some theorems from the previous chapter.

Theorem 85. *We have the following products of means:*

$$(i)\ \mathcal{A}_\lambda(\alpha) \otimes \mathcal{A}_\mu(\alpha) = \mathcal{A}_{\frac{\mu}{1-\lambda+\mu}}(\alpha), \alpha \in CM(J);$$
$$(ii)\ \mathcal{A}_\lambda\left(a^\alpha\right) \otimes \mathcal{A}_{1-\lambda}\left(a^{-\alpha}\right) = \mathcal{A}(\alpha), \alpha \in CM(J), a > 0;$$
$$(iii)\ \mathcal{C}_{p;\lambda} \otimes \mathcal{C}_{2-p;1-\lambda} = \mathcal{A};$$
$$(iv)\ \mathcal{C}_{p;\lambda} \otimes \mathcal{C}_{1-p;1-\lambda} = \mathcal{G};$$
$$(v)\ \mathcal{C}_{p;\lambda} \otimes \mathcal{C}_{-p;1-\lambda} = \mathcal{H};$$
$$(vi)\ \mathcal{S}_{p,q;\lambda} \otimes \mathcal{S}_{-p,-q;1-\lambda} = \mathcal{G};$$
$$(vii)\ \mathcal{S}_{m+p,p;\lambda} \otimes \mathcal{S}_{m-p,-p;1-\lambda} = \mathcal{P}_m;$$
$$(viii)\ \mathcal{D}_{p,q} \otimes \mathcal{D}_{-p,-q} = \mathcal{G};$$
$$(ix) \mathcal{P}_{m;\lambda} \otimes \Pi_1 = \mathcal{P}_{m;\frac{1}{2-\lambda}};$$
$$(x)\ \Pi_2 \otimes \mathcal{P}_{m;\lambda} = \mathcal{P}_{m;\frac{\lambda}{1+\lambda}}.$$

Remark 84. The $\mathcal{AGM}$ can also be represented by

$$\mathcal{A} \otimes \mathcal{G}(a,b) = \frac{\pi}{2} \cdot \left[\int_0^\infty \frac{dt}{\sqrt{(a^2+t^2)(b^2+t^2)}}\right]^{-1}, \tag{3.17}$$

as it was shown in the first chapter. Starting from this, G.D. Song made the following **conjecture:**

$$\mathcal{A}_p \otimes \mathcal{G}_p(a,b) = \pi \cdot \left[\int_{-\infty}^{\infty} \frac{dt}{(a^2+t^2)^p (b^2+t^2)^{1-p}} \right]^{-1}$$

(see Giroux and Rahman, 1982). Although it seems to be a natural hypothesis, it was soon invalidated in Borwein and Borwein (1983) and in Wimp (1985). On the other hand, a similar function,

$$f(a,b) = \int_0^{\infty} \frac{tdt}{\sqrt[3]{(a^3+t^3)(b^3+t^3)^2}},$$

is used in Borwein (1996) to express the common limit of the double sequence given by:

$$a_{n+1} = \frac{a_n + 2b_n}{3}, \; b_{n+1} = \sqrt[3]{b_n \frac{a_n^2 + a_n b_n + b_n^2}{3}}.$$

Namely, for $a_0 = a, \; b_0 = b$, the limit is $f(1,1)/f(a,b)$. In Borwein and Borwein (1991) the limit was expressed by a hypergeometric series.

Remark 85. Making in (3.17) the change of variable $t^2 = s$, we obtain

$$\mathcal{A} \otimes \mathcal{G}(a,b) = \pi \cdot \left[\int_0^{\infty} \frac{ds}{\sqrt{s(a^2+s)(b^2+s)}} \right]^{-1}.$$

This formula leads to the following general integral:

$$R(\alpha; \delta, \delta'; a^2, b^2) = \frac{1}{B(\alpha, \alpha')} \int_0^{\infty} t^{\alpha-1} (t+a^2)^{-\delta} (t+b^2)^{-\delta'} dt,$$

where $\alpha + \alpha' = \delta + \delta', \; Re(\alpha), \; Re(\alpha') > 0$, and B is the beta Euler's function. It was considered in Carlson (1971) and it is related to the Gaussian hypergeometric series by the formula

$$R(\alpha; \delta, \delta'; a^2, b^2) = b^{-2\alpha} F(\alpha; \delta, \delta + \delta'; 1 - a^2/b^2).$$

B.C. Carlson was able to use R for the representation of twelve Gaussian products. What is more remarkable, he made in this way a unified approach to Gauss's algorithm and to Pfaff–Schwab–Borchard's algorithm. Considering the means

$$M_1 = \mathcal{A}, M_2 = \mathcal{G}, M_3 = \mathcal{G}(\mathcal{A}, \Pi_1) \text{ and } M_4 = \mathcal{G}(\mathcal{A}, \Pi_2),$$

he was interested in finding the G-compound means

$$L_{ij} = M_i \otimes M_j \ , \ i, j = 1, 2, 3, 4.$$

Of course

$$L_{12} = \mathcal{A} \otimes \mathcal{G} \text{ and } L_{14} = \mathcal{A} \boxtimes \mathcal{G}.$$

The following results hold generally.

Theorem 86. *For any fixed choice of distinct indices i and j, we have*

$$L_{ij}(a,b) = \left[R(\alpha; \delta, \delta'; a^2, b^2)\right]^{-\frac{1}{2\alpha}},$$

where $(\alpha; \delta, \delta')$ is given by the (i, j)-th entry in the table

$*$	$j=1$	$j=2$	$j=3$	$j=4$
$i=1$	$*$	$\left(\frac{1}{2}, \frac{1}{2}, \frac{1}{2}\right)$	$\left(\frac{1}{4}, \frac{3}{4}, \frac{1}{2}\right)$	$\left(\frac{1}{2}, \frac{1}{2}, 1\right)$
$i=2$	$\left(\frac{1}{2}, \frac{1}{2}, \frac{1}{2}\right)$	$*$	$\left(1, \frac{3}{4}, \frac{1}{2}\right)$	$\left(1, \frac{1}{2}, 1\right)$
$i=3$	$\left(\frac{1}{2}, 1, \frac{1}{2}\right)$	$\left(1, 1, \frac{1}{2}\right)$	$*$	$(1, 1, 1)$
$i=4$	$\left(\frac{1}{4}, \frac{1}{2}, \frac{3}{4}\right)$	$\left(1, \frac{1}{2}, \frac{3}{4}\right)$	$(1, 1, 1)$	$*$

Proof. Let us denote $f_i = M_i(a,b)$. As it is shown in Borwein and Borwein (1987), making in R the substitution

$$t = \frac{s(s + f_2^2)}{s + f_1^2},$$

we get

$$R(\alpha; \delta, \delta'; a^2, b^2) = \frac{1}{B(\alpha, \alpha')} \tag{3.18}$$

$$\cdot \int_0^\infty s^{\alpha'-1}(s + f_1^2)^{\alpha-1}(s + f_2^2)^{\alpha'-1}(s + f_3^2)^{1-2\delta}(s + f_4^2)^{1-2\delta'} ds.$$

If we keep i and j fixed, we can determine values for the parameters for which f_k $(k \neq i, j)$ vanishes in (3.18). They correspond to the entry in the table and the resulting integral R is an invariant for $M_i \otimes M_j$. □

Remark 86. As observed before, L_{12} and L_{14} have been previously identified. An interesting case underlined by B.C. Carlson is

$$L_{34}(a,b) = \sqrt{\frac{a^2 - b^2}{2 \log(a/b)}}, \ (a \neq b).$$

This result can be proved by the invariance principle. Indeed, we have

$$L_{34}(M_3(a,b), M_4(a,b)) = \sqrt{\frac{\mathcal{A}(a,b)\cdot a - \mathcal{A}(a,b)\cdot b}{2\log\sqrt{\frac{\mathcal{A}(a,b)\cdot a}{\mathcal{A}(a,b)\cdot b}}}} = L_{34}(a,b).$$

Another algorithm which has an alternative proof is L_{13}. As B.C. Carlson has shown, it can be expressed by inverse lemniscatic sine. Define

$$arcsl(x) = \int_0^x (1-t^4)^{-1/2}dt,\ x^2 \leq 1$$

and

$$arcslh(x) = \int_0^x (1+t^4)^{-1/2}dt,$$

where $arcsl(x)$ is called the **arc lemniscatic sine** of x and represents the length of arc of the lemniscate from the origin to the point with radial coordinate x. Also, $arcslh(x)$ is called **hyperbolic arc lemniscatic sine** of x but the function does not have a geometric interpretation as simple as $arcsl(x)$. Then

$$L_{13}(a,b) = \begin{cases} \dfrac{\sqrt{a^2-b^2}}{\left(arcsl\ {}^4\sqrt{1-\frac{b^2}{a^2}}\right)^2}, & 0 \leq b < a \\ \dfrac{\sqrt{b^2-a^2}}{\left(arcslh\ {}^4\sqrt{\frac{b^2}{a^2}-1}\right)^2}, & 0 < a < b \end{cases}.$$

Remark 87. From L_{34}, using the method of changing the variables, in Borwein and Borwein (1987) it is deduced that

$$\mathcal{G}(\mathcal{P}_{1/2}, \Pi_1) \otimes \mathcal{G}(\mathcal{P}_{1/2}, \Pi_2) = \mathcal{L}.$$

This result was further generalized in Borwein and Borwein (1993) where it is proved that for every $0 < r < 1$, the following result holds:

$$\mathcal{G}_r(\mathcal{D}_{r,1}, \Pi_1) \otimes \mathcal{G}(\mathcal{D}_{r,1}, \Pi_2) = \mathcal{L}.$$

This follows at once from the invariance principle. For $r = 1/2$ we get the previous result.

Remark 88. The strictly total positivity of the reciprocal of some of the above means was proved in Carlson and Gustafson (1983).

Remark 89. As we saw,

$$M \boxtimes N = M \otimes N(M, \Pi_2)$$

so that we can also apply the above results for the Archimedean products.

3.7 RATE OF CONVERGENCE OF A GAUSSIAN DOUBLE SEQUENCE

Assume that the sequences $(a_n)_{n\geq 0}$ and $(b_n)_{n\geq 0}$ defined by (3.12) have the common limit α. We examine the **rate of convergence** of these sequences to their limit. Let us consider the **errors** of the sequences

$$\delta_n = a_n - \alpha,\ \varepsilon_n = b_n - \alpha,\ n \geq 0.$$

To study them, we use again the Taylor formulas for M and N at the point (α, α). Supposing that the means M and N have continuous partial derivatives up to second order, then we get the relation (3.8) and the similar relation

$$\varepsilon_{n+1} = N_a(\alpha, \alpha)\delta_n + [1 - N_a(\alpha, \alpha)]\,\varepsilon_n + O(\delta_n^2 + \varepsilon_n^2).$$

Using them, the following result was proved in Foster and Phillips (1986).

Theorem 87. *If there is no integer $k \geq 0$ for which $a_k = b_k$ and if*

$$0 < M_a(\alpha, \alpha), N_a(\alpha, \alpha) < 1 \text{ and } M_a(\alpha, \alpha) \neq N_a(\alpha, \alpha)$$

then, as $n \to \infty$,

$$\delta_{n+1} = [M_a(\alpha, \alpha) - N_a(\alpha, \alpha)]\delta_n + O(\delta_n^2)$$

and

$$\varepsilon_{n+1} = [M_a(\alpha, \alpha) - N_a(\alpha, \alpha)]\varepsilon_n + O(\varepsilon_n^2).$$

Supposing that the means M and N have continuous partial derivatives up to the third order, then we have the relation (3.11) and the similar relation

$$\varepsilon_{n+1} = N_a(\alpha, \alpha)\delta_n + [1 - N_a(\alpha, \alpha)]\,\varepsilon_n$$
$$+\frac{1}{2}\,[N_{aa}(\alpha, \alpha)\delta_n - N_{bb}(\alpha, \alpha)\varepsilon_n]\,(\delta_n - \varepsilon_n) + O(|\delta_n|^3 + |\varepsilon_n|^3).$$

Using them in Foster and Phillips (1986) is proved further that:

Theorem 88. *If there is no integer $k \geq 0$ for which $a_k = b_k$ and if*

$$0 < M_a(\alpha,\alpha), N_a(\alpha,\alpha) < 1 \text{ and } M_a(\alpha,\alpha) = N_a(\alpha,\alpha)$$

then, as $n \to \infty$,

$$2[1 - M_a(\alpha,\alpha)]\delta_{n+1} = \{[1 - M_a(\alpha,\alpha)][M_{aa}(\alpha,\alpha) - N_{aa}(\alpha,\alpha)] + M_a(\alpha,\alpha)[M_{bb}(\alpha,\alpha) - N_{bb}(\alpha,\alpha)]\}\delta_n^2 + O\left(|\delta_n|^3\right)$$

and

$$2M_a(\alpha,\alpha)\varepsilon_{n+1} = -\{[1 - M_a(\alpha,\alpha)][M_{aa}(\alpha,\alpha) - N_{aa}(\alpha,\alpha)] + M_a(\alpha,\alpha)[M_{bb}(\alpha,\alpha) - N_{bb}(\alpha,\alpha)]\}\varepsilon_n^2 + O\left(|\varepsilon_n|^3\right)$$

Corollary 63. *If M and N are symmetric means then*

$$\delta_{n+1} = [M_{aa}(\alpha,\alpha) - N_{aa}(\alpha,\alpha)] \cdot \delta_n^2 + O\left(|\delta_n|^3\right)$$

and

$$\varepsilon_{n+1} = -[M_{aa}(\alpha,\alpha) - N_{aa}(\alpha,\alpha)] \cdot \varepsilon_n^2 + O\left(|\varepsilon_n|^3\right).$$

Remark 90. As we saw, most of the "usual" symmetric means have the property (2.30). For such means we have the following consequence.

Corollary 64. *If M and N are symmetric means such that*

$$M_{aa}(\alpha,\alpha) = \frac{c}{\alpha},\ N_{aa}(\alpha,\alpha) = \frac{d}{\alpha},\ c, d \in \mathbb{R},$$

then

$$\delta_{n+1} = \frac{c-d}{\alpha} \cdot \delta_n^2 + O\left(|\delta_n|^3\right)$$

and

$$\varepsilon_{n+1} = -\frac{c-d}{\alpha} \cdot \varepsilon_n^2 + O\left(|\varepsilon_n|^3\right).$$

Remark 91. In the special case of the $\mathcal{AGM}$ we have

$$\delta_{n+1} = \frac{1}{4\alpha} \cdot \delta_n^2 + O\left(|\delta_n|^3\right),\ \varepsilon_{n+1} = -\frac{1}{4\alpha} \cdot \varepsilon_n^2 + O\left(|\varepsilon_n|^3\right).$$

The numerical examples suggest that the convergence is in fact better. This fact was emphasized also in Kellog (1929) by the remark that:

$$a_{n+1} - b_{n+1} = \frac{1 - \sqrt{b_n/a_n}}{1 + \sqrt{b_n/a_n}} \cdot \frac{a_n - b_n}{2}.$$

In Raşa and Toader (1990) it was proved that even in more general circumstances the convergence is indeed much faster.

Theorem 89. *If the means M and N are symmetric and have continuous partial derivatives up to the second order, then for every $a, b \in J$ there are the constants $c > 0$ and $p \in (0,1)$ such that*

$$|a_n - b_n| = |\delta_n - \varepsilon_n| \le c \cdot p^{2^n}, \ \forall n \ge 1.$$

Proof. Let us denote by J' the closed interval determined by a and b. Denote also

$$f(x,y) = M(x,y) - N(x,y)$$

and for a fixed $x \in J'$, $g(y) = f(x,y)$. We have $g(x) = f(x,x) = 0$ and $g'(x) = f_y(x,x) = 0$. Thus the Taylor expansion of g in x yields a z between x and y such that $g(y) = (y-x)^2 \cdot g''(z)/2$ or

$$f(x,y) = (y-x)^2 \cdot f_{yy}(z,z)/2.$$

Denoting

$$K = \max_{z \in J'} \left| f_{yy}(z,z) \right| /2,$$

it follows that

$$|M(x,y) - N(x,y)| \le K \cdot (y-x)^2$$

and so

$$\left| a_{j+1} - b_{j+1} \right| \le K \cdot \left(a_j - b_j \right)^2.$$

Taking the 2^{n-j-1}-th power and multiplying the inequalities for $j = i, i+1, ..., n-1$, we get

$$|a_n - b_n| \le [K \cdot (a_i - b_i)]^{2^{n-i}} /K.$$

As $(a_n)_{n \ge 0}$ and $(b_n)_{n \ge 0}$ have a common limit, we find i_0 such that

$$p = \left[K \cdot \left(a_{i_0} - b_{i_0} \right) \right]^{2^{-i_0}} < 1$$

thus

$$|a_n - b_n| \le p^{2^n} /K, \ \forall n \ge i_0.$$

For

$$c = \max \left\{ 1/K, |a_n - b_n| /p^{2^n}, n = 1, ..., i_0 - 1 \right\}$$

we obtain the result. □

3.8 COMPARISON OF COMPOUND MEANS

A result about the comparison of the compound means $M \boxplus N$ and $M \otimes N$ was given in Toader (1991).

Theorem 90. *If the means M and N are isotone and M is A-compoundable and G-compoundable with N, then the following relation*

$$M \otimes N \prec M \boxplus N$$

holds.

Proof. For two initial values $a, b \in J$ define two pairs of sequences: $(a_n)_{n\geq 0}$ and $(b_n)_{n\geq 0}$ by

$$a_{n+1} = M(a_n, b_n) \text{ and } b_{n+1} = N(a_n, b_n)\ ,\ n \geq 0$$

and $\left(a_n'\right)_{n\geq 0}$ and $\left(b_n'\right)_{n\geq 0}$ by

$$a_{n+1}' = M(a_n', b_n') \text{ and } b_{n+1}' = N(a_{n+1}', b_n'),\ n \geq 0,$$

where $a_0 = a_0' = a,\ b_0 = b_0' = b$. If $a < b$ we can prove by induction that

$$a_n \leq a_n' \text{ and } b_n \leq b_n' \text{ for all } n \geq 0.$$

Indeed, the relations hold with equality for $n = 0$. Assuming them valid for n, we have

$$a_{n+1} = M(a_n, b_n) \leq M(a_n', b_n) \leq M(a_n', b_n') = a_{n+1}',$$

by the isotony of M. Then, as it was proved in (3.5), $a_n' \leq a_{n+1}'$, thus

$$b_{n+1} = N(a_n, b_n) \leq N(a_n', b_n') \leq N(a_{n+1}', b_n') = b_{n+1}'.$$

Therefore

$$M \otimes N(a,b) = \lim_{n\to\infty} a_n \leq \lim_{n\to\infty} a_n' = M \boxplus N(a,b).$$

For $a > b$ we get $M \otimes N(a,b) \geq M \boxplus N(a,b)$, which completes the proof of the relation. □

Example 10. The following inequality

$$\mathcal{A} \otimes \mathcal{G} \prec \mathcal{A} \boxtimes \mathcal{G}$$

holds. Thus

$$\frac{\pi}{2}\cdot\left[\int_0^{\pi/2}\frac{d\theta}{\sqrt{a^2\cos^2\theta+b^2\sin^2\theta}}\right]^{-1}\le\frac{\sqrt{b^2-a^2}}{\arccos(a/b)},\quad \text{for } 0<a<b$$

and

$$\frac{\pi}{2}\cdot\left[\int_0^{\pi/2}\frac{d\theta}{\sqrt{a^2\cos^2\theta+b^2\sin^2\theta}}\right]^{-1}\ge\frac{\sqrt{a^2-b^2}}{\arg\cosh(a/b)},\quad \text{for } 0<b<a.$$

In Borwein and Borwein (1993) there is proved a Comparison Lemma which generalizes the Invariance Principle and allows proving some inequalities for G-compound means. We present here the following variant given in Costin and Toader (2006a). Denote

$$\Delta_J^+=\{(a,b)\in J\times J,\ a\ge b\}.$$

Theorem 91. *Let the mean M be G-compoundable with the mean N and $M\ge N$ on Δ_J^+. If the continuous pre-mean P has the property*

$$P(M,N)\le P \text{ on } \Delta_J^+, \tag{3.19}$$

then

$$M\otimes N\le P \text{ on } \Delta_J^+.$$

Proof. For $(a,b)\in\Delta_J^+$ define the sequences $(a_n)_{n\ge0}$ and $(b_n)_{n\ge0}$ by

$$a_{n+1}=M(a_n,b_n) \text{ and } b_{n+1}=N(a_n,b_n),\ n\ge0,$$

where $a_0=a,\ b_0=b$. As $M\ge N$, we have

$$a_n\ge a_{n+1}\ge b_{n+1}\ge b_n \text{ for all } n\ge0,$$

thus $(a_n,b_n)\in\Delta_J^+$ for all $n\ge0$. From (3.19) we deduce that

$$P(a_{n+1},b_{n+1})\le P(a_n,b_n)\le P(a,b) \text{ for all } n\ge0.$$

Passing to the limit, we reach our conclusion because P is a pre-mean. □

Similar results can be obtained on

$$\Delta_J^-=\{(a,b)\in J\times J,\ a\le b\}$$

and/or for $M\le N$.

The Comparison Lemma from Borwein and Borwein (1993) can now be obtained as follows. Denote

$$\Delta_\varepsilon^+ = \{(a,b) \in \mathbb{R}_+^2;\ 1 > b/a > \varepsilon \geq 0\}.$$

Corollary 65. *Let M, N be strict continuous means on Δ_ε^+. If the continuous pre-mean P has the property*

$$P(M,N) \leq P \text{ on } \Delta_\varepsilon^+,$$

then

$$M \otimes N \leq P \text{ on } \Delta_\varepsilon^+.$$

Proof. For $(a,b) \in \Delta_\varepsilon^+$ we have $\Delta_{[b,a]}^+ \subset \Delta_\varepsilon^+$, thus we can apply the previous theorem obtaining

$$M \otimes N(a,b) \leq P(a,b).$$

□

Using these results, some consequences were proved in Costin and Toader (2006a).

Theorem 92. *If the following conditions are satisfied on Δ_J^+: (i) M is G-compoundable with N and M' is G-compoundable with N'; (ii) $M < M'$ and $N < N'$; (iii) the means M' and N' are isotone; then the following relation*

$$M \otimes N \leq M' \otimes N'$$

holds on Δ_J^+.

Proof. If we denote $M' \otimes N' = P'$, we deduce that P' is isotone. Thus, for $(a,b) \in \Delta_J^+$, using the invariance principle, we have

$$P'(M(a,b), N(a,b)) \leq P'(M'(a,b), N'(a,b)) = P'(a,b),$$

which gives the desired result. □

Theorem 93. *If the following conditions are satisfied on Δ_J^+: (i) M is G-compoundable with N and M' is G-compoundable with N'; (ii) $M < M'$ and $N < N'$; (iii) the means M and N are isotone; then the following relation*

$$M \otimes N \leq M' \otimes N'$$

holds on Δ_J^+.

Proof. We use the invariance principle and the isotony of $M \otimes N = P$. □

Corollary 66. *If the following conditions are satisfied: (i) M is G-compoundable with N and M' is G-compoundable with N'; (ii) $M < M'$ and $N < N'$; (iii) the means M' and N' (or M and N) are isotone; then the following relation*

$$M \otimes N \leq M' \otimes N'$$

holds.

Proof. As $J \times J = \Delta_J^+ \cup \Delta_J^-$, we apply the previous theorems separately on Δ_J^+ and on Δ_J^-. □

Apparently we cannot prove in the same way the following:

Theorem 94. *If the following conditions are satisfied: (i) M is G-compoundable with N and M' is G-compoundable with N'; (ii) $M < M'$ and $N < N'$; (iii) at least one of the means M, M' and one of the means N, N' is isotone; then the following relation*

$$M \otimes N \leq M' \otimes N'$$

holds.

Proof. For two initial values $a, b \in J$ define two pairs of sequences: $(a_n)_{n \geq 0}$ and $(b_n)_{n \geq 0}$ by

$$a_{n+1} = M(a_n, b_n) \text{ and } b_{n+1} = N(a_n, b_n),\ n \geq 0$$

and $\left(a_n'\right)_{n \geq 0}$ and $\left(b_n'\right)_{n \geq 0}$ by

$$a_{n+1}' = M'(a_n', b_n') \text{ and } b_{n+1}' = N'(a_n', b_n'),\ n \geq 0,$$

where $a_0 = a_0' = a, b_0 = b_0' = b$. We can prove by induction that

$$a_n \leq a_n' \text{ and } b_n \leq b_n' \text{ for all } n \geq 0.$$

Assuming them valid for n and M to be isotone, we have

$$a_{n+1} = M(a_n, b_n) \leq M(a_n', b_n') \leq M'(a_n', b_n') = a_{n+1}'.$$

If M' is isotone, the same relation follows by a small change:

$$a_{n+1} = M(a_n, b_n) \leq M'(a_n, b_n) \leq M'(a_n', b_n') = a_{n+1}'.$$

Similarly it is proved that $b_{n+1} \leq b_{n+1}'$. Thus

$$M \otimes N(a, b) = \lim_{n \to \infty} a_n \leq \lim_{n \to \infty} a_n' = M' \otimes N'(a, b).$$ □

Example 11. We have

$$\mathcal{A}(\mathcal{G},\Pi_1)\otimes\mathcal{A}(\mathcal{G},\Pi_2)\le\mathcal{G}(\mathcal{A},\Pi_1)\otimes\mathcal{G}(\mathcal{A},\Pi_2).$$

Indeed, all these means are isotone and

$$\mathcal{A}(\mathcal{G},\Pi_1)\le\mathcal{G}(\mathcal{A},\Pi_1),\ \mathcal{A}(\mathcal{G},\Pi_2)\le\mathcal{G}(\mathcal{A},\Pi_2),$$

both these inequalities being equivalent to the known inequality $\mathcal{P}_{1/2}\le\mathcal{P}_1$. As we saw, B.C. Carlson proved that

$$\mathcal{A}(\mathcal{G},\Pi_1)\otimes\mathcal{A}(\mathcal{G},\Pi_2)=\mathcal{L}$$

and

$$\mathcal{G}(\mathcal{A},\Pi_1)\otimes\mathcal{G}(\mathcal{A},\Pi_2)=\mathcal{G}(\mathcal{A},\mathcal{L}).$$

So the inequality between the given G-compound means gives

$$\mathcal{L}\le\mathcal{A},$$

which is well known.

Remark 92. In Carlson and Vuorinen (1991) given is the inequality

$$\mathcal{L}\le\mathcal{A}\otimes\mathcal{G},\tag{3.20}$$

while in Vamanamurthy and Vuorinen (1994) proved is the inequality

$$\mathcal{A}\otimes\mathcal{G}\le\mathcal{G}(\mathcal{A},\mathcal{L}).\tag{3.21}$$

These inequalities are equivalent to the following inequalities among G-compound means:

$$\mathcal{A}(\mathcal{G},\Pi_1)\otimes\mathcal{A}(\mathcal{G},\Pi_2)\le\mathcal{A}\otimes\mathcal{G}\le\mathcal{G}(\mathcal{A},\Pi_1)\otimes\mathcal{G}(\mathcal{A},\Pi_2).$$

For their proof, we cannot use the previous theorems because the means are not comparable. As they are weakly comparable, in Costin and Toader (2006a) results based on such hypotheses are established.

Corollary 67. *If the following conditions are satisfied: (i) M is G-compoundable with N and M' is G-compoundable with N'; (ii) $M\prec M'$ and $N\prec N'$; (iii) the means M' and N' (or M and N) are isotone; then the following relation*

$$M\otimes N\prec M'\otimes N'$$

holds.

Theorem 95. *If the following conditions are satisfied: (i) at least one of the means M, M' and one of the means N, N' is isotone; (ii) M is A-compoundable with N and M' is A-compoundable with N'; (iii) $M < M'$ and $N < N'$; then the following relation*

$$M \boxtimes N \leq M' \boxtimes N'$$

holds.

Remark 93. Unfortunately, the inequalities (3.20) and (3.21) cannot be deduced from this theorem. The first inequality was proved in Braden et al. (1992) and Todd (1992). We present here the proof from Borwein and Borwein (1993). First of all, using the homogeneity of the means, the inequality $\mathcal{L} \leq \mathcal{A} \otimes \mathcal{G}$ is equivalent to

$$\mathcal{L}(1,x) \leq \mathcal{A} \otimes \mathcal{G}(1,x),\ x \in (0,1).$$

Using the comparison lemma, we have to prove that

$$\mathcal{L}(1,x) \leq \mathcal{L}(\mathcal{A}, \mathcal{G})(1,x),\ x \in (0,1),$$

or

$$(1+\sqrt{x}) \cdot \ln \frac{1+x}{2} < \sqrt{x} \cdot \ln x,\ x \in (0,1).$$

Substituting $x = y^2$, it becomes

$$\frac{1+y}{2} \cdot \ln \frac{1+y^2}{2} < y \cdot \ln y,\ y \in (0,1).$$

Define the function $g(y) = y \cdot \ln y$. It is convex and $g(1) = 0$. As $(1+y^2)/2 > [(1+y)/2]^2$, it follows that

$$\frac{1+y}{4} \cdot \ln \frac{1+y^2}{2} < \frac{1+y}{2} \cdot \ln \frac{1+y}{2} = g\left(\frac{1+y}{2}\right) < \frac{g(y)}{2} + \frac{g(1)}{2} = \frac{g(y)}{2}.$$

In the same paper the following inequality is also given:

$$\mathcal{A} \otimes \mathcal{G} < \mathcal{L}_{3/2}.$$

As above, it suffices to show that

$$\mathcal{L}_{3/2}(1,x) > \mathcal{L}_{3/2}(\mathcal{A}, \mathcal{G})(1,x),\ x \in (0,1).$$

Replacing x by $x^{2/3}$, this becomes

$$\mathcal{L}(1,x) > \mathcal{L}(\mathcal{P}_{2/3}, \mathcal{G})(1,x),\ x \in (0,1).$$

Written out explicitly, this becomes

$$\frac{\ln x}{x-1} < \frac{\ln\left[\left(\frac{1+x^{2/3}}{2}\right)^{3/2}/\sqrt{x}\right]}{\left(\frac{1+x^{2/3}}{2}\right)^{3/2} - \sqrt{x}}, \ x \in (0,1).$$

It was established by the authors only "by a somewhat unsatisfactory computational route" which was only sketched. An explanation of the difficulty in proving it was found in the fact that the first four terms of the Taylor series, for $x = 1$, on each side of the inequality coincide.

Remark 94. In Vamanamurthy and Vuorinen (1994) the following inequality is also proved:

$$\mathcal{A} \otimes \mathcal{G} \leq \mathcal{P}_{1/2}.$$

The original proof is quite complicated. We give here a simpler one, which is in the lines of other results of this paragraph. In fact, we shall give a complete result taken from Costin and Toader (2006a).

Theorem 96. *The $\mathcal{AGM}$ verifies the following inequalities:*

$$\mathcal{P}_0 = \mathcal{G} < \mathcal{A} \otimes \mathcal{G} < \mathcal{P}_{1/2}. \tag{3.22}$$

The inequalities cannot be improved, thus $\mathcal{A} \otimes \mathcal{G}$ is not comparable with $\mathcal{P}_q$ for $0 < q < 1/2$.

Proof. As we already saw, the first inequality follows from the definition of the $\mathcal{AGM}$. The second inequality is equivalent to

$$\mathcal{P}_{1/2}(\mathcal{A}, \mathcal{G}) < \mathcal{P}_{1/2},$$

thus

$$\sqrt{\frac{1+x}{2}} + \sqrt{\sqrt{x}} < 2 \cdot \frac{1+\sqrt{x}}{2}, \text{ for all } x \in (0,1).$$

Making the change of variable $x^{1/4} = y$, we obtain

$$\sqrt{\frac{1+y^4}{2}} + y < 1 + y^2, \text{ for all } y \in (0,1),$$

which, after some obvious manipulations, becomes $(y-1)^4 > 0$. Let us prove now that the inequalities cannot be improved. Assume that

$$\mathcal{A} \otimes \mathcal{G} < \mathcal{P}_q. \tag{3.23}$$

According to the Comparison Lemma, we have

$$\mathcal{A}^q + \mathcal{G}^q < 2 \cdot \mathcal{P}_q^q,$$

or

$$\left(\frac{1+x}{2}\right)^q + x^{\frac{q}{2}} < 2 \cdot \frac{1+x^q}{2}, \; x \in (0,1).$$

Denoting $x = 1 - y$, this becomes

$$\left(1 - \frac{y}{2}\right)^q + (1-y)^{\frac{q}{2}} < 1 + (1-y)^q, \; y \in (0,1).$$

Using the binomial series, we get

$$1 - q \cdot \frac{y}{2} + \frac{q(q-1)}{2} \cdot \left(\frac{y}{2}\right)^2 - \ldots + 1 - \frac{q}{2} \cdot y + \frac{q(q-2)}{8} \cdot y^2 + \ldots$$
$$< 1 + 1 - q \cdot y + \frac{q(q-1)}{2} \cdot y^2 - \ldots, \; y \in (0,1),$$

thus

$$\frac{q(1-2q)}{8} \cdot y^2 + \ldots < 0, \; y \in (0,1). \tag{3.24}$$

Taking into account (3.22), the inequality (3.23) cannot be valid for $q < 0$, thus $q \geq 1/2$. Similarly, the reversed inequality of (3.23) implies the reversed inequality of (3.24). This time $q < 1/2$, which implies $q \leq 0$. □

Corollary 68. *We have the following evaluation for the* $\mathcal{AGM}$*:*

$$\mathcal{G} < \mathcal{L} < \mathcal{A} \otimes \mathcal{G} < \mathcal{P}_{1/2} < \mathcal{P}_{2/3} < \mathcal{I} < \mathcal{P}_{\ln 2} < \mathcal{A}.$$

Remark 95. In Vamanamurthy and Vuorinen (1994), using some complicated inequalities with elliptic complete integrals, it is proved that

$$\mathcal{A} \otimes \mathcal{G} < \frac{\pi}{2} \cdot \mathcal{L}.$$

So we have the logarithmic evaluation for the $\mathcal{AGM}$

$$\mathcal{L} < \mathcal{A} \otimes \mathcal{G} < \frac{\pi}{2} \cdot \mathcal{L},$$

together with that given before:

$$\mathcal{L} < \mathcal{A} \otimes \mathcal{G} < \mathcal{L}_{3/2}.$$

Other inequalities can be found in Sándor (1995, 1996, 2001), or Yang et al. (2014).

3.9 THE SCHWAB–BORCHARDT MEAN

The mean $\mathcal{A} \boxtimes \mathcal{G}$ was named **Schwab–Borchardt mean** (see Neuman and Sándor, 2003) and we denote it by $\mathcal{B}$.

From (3.16) we deduce the invariance formula

$$\mathcal{B} = \mathcal{B}(\mathcal{A}, \mathcal{G}(\mathcal{A}, \Pi_2)).$$

Using it, in Neuman and Sándor (2003) the following is given:

Lemma 18. *If $a > b$, then*

$$\mathcal{B}(a, b) < \mathcal{B}(b, a).$$

In the same paper it is proved that

$$\mathcal{B}(\mathcal{A}, \mathcal{G}) = \mathcal{L} \tag{3.25}$$

$$\mathcal{B}(\mathcal{G}, \mathcal{A}) = \mathcal{R} \tag{3.26}$$

$$\mathcal{B}(\mathcal{A}, \mathcal{Q}) = \mathcal{T} \tag{3.27}$$

and

$$\mathcal{B}(\mathcal{Q}, \mathcal{A}) = \mathcal{N} \tag{3.28}$$

where

$$\mathcal{R}(a, b) = \frac{a - b}{2 \arcsin\left(\dfrac{a - b}{a + b}\right)} \tag{3.29}$$

is the **first Seiffert mean**, introduced in Seiffert (1993),

$$\mathcal{T}(a, b) = \frac{a - b}{2 \arctan\left(\dfrac{a - b}{a + b}\right)} \tag{3.30}$$

is the **second Seiffert mean** introduced in Seiffert (1995), and

$$\mathcal{N}(a, b) = \frac{a - b}{2\operatorname{arcsinh}\left(\dfrac{a - b}{a + b}\right)}$$

is the **Neuman–Sándor mean** defined in Neuman and Sándor (2003).

As the logarithmic mean can be given by

$$\mathcal{L}(a, b) = \frac{a - b}{2\operatorname{arctanh}\left(\dfrac{a - b}{a + b}\right)},$$

we see that these four means $\mathcal{L}$, $\mathcal{R}$, $\mathcal{T}$, and $\mathcal{N}$, are very similar. They together are called **Seiffert-type means**.

Using the monotonicity of the mean $\mathcal{B}$ in its arguments, Lemma 18, and the inequalities $\mathcal{G} < \mathcal{A} < \mathcal{Q}$, in Neuman and Sándor (2003) it is proved that

$$\mathcal{G} = \mathcal{B}(\mathcal{G}, \mathcal{G}) < \mathcal{B}(\mathcal{A}, \mathcal{G}) < \mathcal{B}(\mathcal{G}, \mathcal{A}) < \mathcal{B}(\mathcal{A}, \mathcal{A})$$
$$= \mathcal{A} < \mathcal{B}(\mathcal{Q}, \mathcal{A}) < \mathcal{B}(\mathcal{A}, \mathcal{Q}) < \mathcal{B}(\mathcal{Q}, \mathcal{Q}) = \mathcal{Q}$$

thus

$$\mathcal{G} < \mathcal{L} < \mathcal{R} < \mathcal{A} < \mathcal{N} < \mathcal{T} < \mathcal{Q}.$$

The first three inequalities are known from Carlson (1972), Seiffert (1993) and Seiffert (1995a), while the last inequality appears in Seiffert (1995). We have referred also to Lin (1974) giving optimal evaluations of the logarithmic mean by power means (2.11). In the last ten years many papers were devoted to these means. For instance, in Jagers (1994) the following estimations are proven:

$$\mathcal{P}_{1/3} < \mathcal{R} < \mathcal{P}_{2/3},$$

while in Seiffert (1995) it is given that

$$\mathcal{P}_1 < \mathcal{T} < \mathcal{P}_2.$$

Optimal estimations for the first Seiffert mean are given in Hästö (2004):

$$\mathcal{P}_{\frac{\ln 2}{\ln \pi}} < \mathcal{R} < \mathcal{P}_{2/3}.$$

We consider also the following.

Definition 23. Two means $\mathcal{M} < \mathcal{N}$ can be separated by a family of means F_p, $p \in \mathbb{R}$ if there is an index p such that $\mathcal{M} < F_p < \mathcal{N}$.

The first separation of the means $\mathcal{N} < \mathcal{T}$ by power means was given in Costin and Toader (2012):

$$\mathcal{N} < \mathcal{P}_{3/2} < \mathcal{T}.$$

The proof is based on Theorem 21, the relations (3.27) and (3.28), and the double-sided inequality

$$\mathcal{G}_{1/3} < \mathcal{B} < \mathcal{A}_{1/3},$$

proved in Neuman and Sándor (2003). We have

$$\mathcal{N} = \mathcal{B}(\mathcal{Q}, \mathcal{A}) < \frac{\mathcal{Q} + 2\mathcal{A}}{3} < \mathcal{D}_{3,1} < \mathcal{D}_{3,3/2}$$

$$= \mathcal{P}_{3/2} < \mathcal{D}_{4,1} = \sqrt[3]{\mathcal{A} \cdot \mathcal{Q}_2^2} < \mathcal{B}(\mathcal{A}, \mathcal{Q}_2) = \mathcal{T}.$$

Using the above results, in Costin and Toader (2012) the following nice chain of inequalities was obtained:

$$\mathcal{P}_0 < \mathcal{L} < \mathcal{P}_{1/2} < \mathcal{R} < \mathcal{P}_1 < \mathcal{N} < \mathcal{P}_{3/2} < \mathcal{T} < \mathcal{P}_2.$$

Including in this chain of inequalities the identric mean as in (2.12), in Costin and Toader (2012a) another chain of means is obtained, in which some Seiffert-type means are separated by power means with equidistant indices:

$$\mathcal{P}_0 < \mathcal{L} < \mathcal{P}_{1/3} < \mathcal{R} < \mathcal{P}_{2/3} < \mathcal{I} < \mathcal{P}_1 < \mathcal{N} < \mathcal{P}_{4/3} < \mathcal{T} < \mathcal{P}_{5/3} < \mathcal{P}_{6/3} < \mathcal{J}.$$

Finally, optimal estimations by power means were obtained independently in Costin and Toader (2013) and Yang (2012) for the second Seiffert mean

$$\mathcal{P}_{\ln 2/\ln(\pi/2)} < \mathcal{T} < \mathcal{P}_{5/3},$$

and in Costin and Toader (2013), Yang (2012a), and Chu and Long (2013) for the Neuman–Sándor mean:

$$\mathcal{P}_{\ln 2/\ln(\ln(3+2\sqrt{2}))} < \mathcal{N} < \mathcal{P}_{4/3}.$$

So the optimal estimations by power means can be ordered as:

$$\mathcal{P}_0 < \mathcal{L} < \mathcal{P}_{1/3} < \mathcal{P}_{\ln/\ln\pi} < \mathcal{R} < \mathcal{P}_{2/3} < \mathcal{I} < \mathcal{P}_{\ln 2} < \mathcal{P}_{\ln 2/\ln(\ln(3+2\sqrt{2}))}$$
$$< \mathcal{N} < \mathcal{P}_{4/3} < \mathcal{P}_{\ln 2/\ln(\pi/2)} < \mathcal{T} < \mathcal{P}_{5/3} < \mathcal{P}_2 < \mathcal{J} < \mathcal{P}_\infty.$$

As we saw in Theorem 18, the logarithmic mean was estimated also by other families of means. Using Lehmer means, in Alzer (1993) the lower estimation for the identric mean was given:

$$\mathcal{C}_{5/6} < \mathcal{I},$$

in Wang et al. (2010) the Seiffert means were estimated:

$$\mathcal{C}_{5/6} < \mathcal{R} < \mathcal{C}_1$$

and

$$\mathcal{C}_1 < \mathcal{T} < \mathcal{C}_{4/3},$$

while in Costin and Toader (2014) the optimal estimations are given:

$$\mathcal{C}_1 < \mathcal{N} < \mathcal{C}_{7/6},$$

$$\mathcal{C}_{3/2} < \mathcal{J} < \mathcal{C}_2,$$

and it is remarked that the above optimal estimations can only be ordered in two chains:

$$\mathcal{C}_{2/3} < \mathcal{L} < \mathcal{C}_1 < \mathcal{N} < \mathcal{C}_{7/6} < \mathcal{C}_{3/2} < \mathcal{J} < \mathcal{C}_2$$

and

$$\mathcal{C}_{5/6} < \mathcal{R} < \mathcal{T} < \mathcal{C}_1 < \mathcal{J} < \mathcal{C}_{4/3},$$

as the means $\mathcal{L} < \mathcal{R} < \mathcal{I}$ and $\mathcal{N} < \mathcal{T}$ cannot be separated by Lehmer means.

Similarly, in Costin and Toader (2014a) optimal estimations by the special Gini means S_p are determined. They can be given in two chains of optimal inequalities

$$\mathcal{S}_{1/3} < \mathcal{L} < \mathcal{S}_1 < \mathcal{N} < \mathcal{S}_{4/3}$$

and

$$\mathcal{S}_{2/3} < \mathcal{R} < \mathcal{I} < \mathcal{S}_1 < \mathcal{N} < \mathcal{T} < \mathcal{S}_{5/3} < \mathcal{J} < \mathcal{S}_2.$$

The pairs of means $\mathcal{L} < \mathcal{P}$, $\mathcal{P} < \mathcal{I}$, and $\mathcal{N} < \mathcal{T}$ cannot be separated by these special Gini means. Also, as in (2.13), the optimal estimation of each of the means $\mathcal{R}$, $\mathcal{I}$, $\mathcal{N}$, $\mathcal{T}$, and $\mathcal{J}$ can be ordered as follows:

$$\begin{gathered}
\mathcal{P}_{\ln 2/\ln\pi},\ \mathcal{S}_{2/3} < \mathcal{C}_{5/6} < \mathcal{R} < \mathcal{P}_{2/3} < \mathcal{C}_1 = \mathcal{S}_1 = \mathcal{A} \\
\mathcal{S}_{2/3},\ \mathcal{C}_{5/6} < \mathcal{P}_{2/3} < \mathcal{I} < \mathcal{P}_{\ln 2} < \mathcal{C}_1 = \mathcal{S}_1 = \mathcal{A} \\
\mathcal{A} = \mathcal{C}_1 = \mathcal{S}_1 < \mathcal{P}_{\ln 2/\ln(\ln(2+2\sqrt{2}))} < \mathcal{N} < \mathcal{P}_{4/3} < \mathcal{C}_{7/6} < \mathcal{S}_{4/3} \\
\mathcal{A} = \mathcal{C}_1 = \mathcal{S}_1 = \mathcal{P}_{\ln 2/\ln(\pi/2)} < \mathcal{T} < \mathcal{P}_{5/3} < \mathcal{C}_{4/3} < \mathcal{S}_{5/3} \\
\mathcal{P}_2 < \mathcal{C}_{3/2} < \mathcal{S}_2 = \mathcal{J} = \mathcal{S}_2 < \mathcal{C}_2 < \mathcal{P}_\infty.
\end{gathered}$$

We have to remark that $\mathcal{S}_{2/3}$ is not comparable with $\mathcal{P}_{\ln 2/\ln\pi}$ and $\mathcal{C}_{5/6}$.

3.10 SEIFFERT-LIKE MEANS

Taking into account the representation of the Seiffert means $\mathcal{R}$ and $\mathcal{T}$ by (3.29) and (3.30) respectively, in Toader (1998b, 1999) the **generalized Seiffert means** $\mathcal{R}_{M,p}$ and $\mathcal{T}_{M,p}$ are defined by:

$$\mathcal{R}_{M,p}(a,b) = \frac{a-b}{p \arcsin \dfrac{a-b}{p \cdot M(a,b)}}$$

respectively

$$\mathcal{T}_{M,p}(a,b) = \frac{a-b}{p \arctan \dfrac{a-b}{p \cdot M(a,b)}}$$

where M is a mean and p a positive real number satisfying some conditions, thus

$$\mathcal{R}_{M,p} = M \cdot \frac{u}{\arcsin u}, \quad \mathcal{T}_{M,p} = M \cdot \frac{u}{\arctan u},$$

with $u = \dfrac{a-b}{p \cdot M(a,b)}$.

A special case $\mathcal{R}_p = \mathcal{R}_{\mathcal{A},\frac{1}{p}}$, $\mathcal{T}_p = \mathcal{T}_{\mathcal{A},\frac{1}{p}}$ was redefined and used in Chu et al. (2011) and a similar one in Neuman (2013). Other generalizations of the Seiffert means were defined in Neuman (2014) using the Schwab–Borchardt mean as in (3.25)–(3.28).

A general construction of **Seiffert-like means** was developed in Witkowski (2013). For positive a, b and a positive function $f : (0,1) \to \mathbb{R}$, let

$$S_f(a,b) = \begin{cases} \dfrac{|a-b|}{2f\left(\dfrac{|a-b|}{a+b}\right)}, & a \neq b \\ a, & a = b. \end{cases}$$

The author was interested under what assumptions on f, the function S_f is a mean. He found the following answer. Denote by S the set of functions $f : (0,1) \to \mathbb{R}$ satisfying

$$\frac{z}{1+z} \leq f(z) \leq \frac{z}{1-z}. \tag{3.31}$$

They are called **Seiffert functions**. Denote also by H the set of homogeneous symmetric means.

Theorem 97. *The mapping $f \to S_f$ is a one-to-one correspondence between S and H.*

If $M \in H$ and $a < b$, then

$$M(a,b) = \frac{a+b}{2} \cdot M\left(\frac{a+b-(b-a)}{a+b}, \frac{a+b+(b-a)}{a+b}\right)$$
$$= \frac{b-a}{2\dfrac{z}{M(1+z,1-z)}},$$

where $z = \dfrac{b-a}{b+a}$, *thus*

$$f_M(z) = \frac{z}{M(1+z, 1-z)}$$

is the Seiffert function of the mean M.

For instance,

$$f_\wedge(z) = \frac{z}{1-z}, \quad f_\vee(z) = \frac{z}{1+z}.$$

Writing the condition (3.31) as

$$-1 \le \frac{1}{f(z)} - \frac{1}{z} \le 1$$

we see that

$$d_S(f, g) = \sup_{0<z<1} \left| \frac{1}{f(z)} - \frac{1}{g(z)} \right|$$

defines a metric on S. It induces a metric on H as

$$d_H(M, N) = d_S(f_M, f_N) = \sup_{0<z<1} \left| \frac{1}{f_M(z)} - \frac{1}{f_N(z)} \right|$$

$$= \sup_{a \ne b} \frac{2}{|a-b|} \left| \frac{|a-b|}{2f_M(z)} - \frac{|a-b|}{2f_N(z)} \right| = 2 \sup_{a \ne b} \left| \frac{M(a,b) - N(a,b)}{a-b} \right|$$

which is analogous to (2.25).

It can also be used to prove that if $M, N \in H$ are such that $d_H(M, N) < 2$, then there exists an unique mean $P \in H$ which is (M, N)-invariant.

3.11 DOUBLE SEQUENCES WITH PRE-MEANS

As it was shown in Toader and Toader (2007), the arithmetic–geometric process of Gauss can be generalized as follows. Let us consider two functions M and N defined on a set D and let $(a, b) \in D$ be an initial point.

Definition 24. If the pair of sequences $(a_n)_{n \ge 0}$ and $(b_n)_{n \ge 0}$ can be defined by

$$a_{n+1} = M(a_n, b_n) \text{ and } b_{n+1} = N(a_n, b_n)$$

for each $n \ge 0$, where $a_0 = a,\ b_0 = b$, then it is called a **Gaussian double sequence**. The function M is **compoundable in the sense of Gauss** (or **G-compoundable**) with the function N if the sequences $(a_n)_{n \ge 0}$ and $(b_n)_{n \ge 0}$ are defined and convergent to a common limit $M \otimes N(a, b)$ for each

$(a, b) \in D$. In this case $M \otimes N$ is called the **Gaussian compound function** (or **G-compound function**).

Remark 96. If M and N are G-compoundable means, then $M \otimes N$ is also a mean called the **G-compound mean.**

In the case of means, the method of searching G-compound functions is based generally on the **invariance principle**. In Toader and Toader (2007) the following generalized invariance principle was proved.

Theorem 98. *Let P be a continuous pre-mean on D and M and N be two functions on D such that N is the P-complementary of M. If the sequences $(a_n)_{n\geq 0}$ and $(b_n)_{n\geq 0}$ defined by*

$$a_{n+1} = M(a_n, b_n) \text{ and } b_{n+1} = N(a_n, b_n),\ n \geq 0,$$

are convergent to a common limit L denoted as $M \otimes N(a_0, b_0)$, then this limit is

$$M \otimes N(a_0, b_0) = P(a_0, b_0).$$

Proof. As N is the P-complementary of M, we have

$$P(M(a_n, b_n), N(a_n, b_n)) = P(a_n, b_n), \forall n \geq 0,$$

thus

$$P(a_{n+1}, b_{n+1}) = P(a_n, b_n), \forall n \geq 0.$$

But this also means that

$$P(a_0, b_0) = P(a_n, b_n), \forall n \geq 0.$$

Finally, as P is a continuous pre-mean, passing to the limit we get

$$P(a_0, b_0) = P(L, L) = L,$$

which proves the result. □

It is natural to study the following

Problem. If N is the P-complementary of M but M, N, or P are not means, are the sequences $(a_n)_{n\geq 0}$ and $(b_n)_{n\geq 0}$ convergent?

The answer can be positive as it is shown in the following:

Example 12. We have ${}^{\mathcal{G}_{4/5}}\mathcal{G}_{5/8} = \mathcal{G}_{3/2}$, where $\mathcal{G}_{3/2}$ is not a mean. Take $a_0 = 10^5, b_0 = 1$ and

$$a_{n+1} = \mathcal{G}_{5/8}(a_n, b_n), b_{n+1} = \mathcal{G}_{3/2}(a_n, b_n),\ n \geq 0.$$

Though some of the first terms take values out of the interval $[b_0, a_0]$ like

$$b_1 \approx 3.1 \cdot 10^7, b_3 \approx 4.7 \cdot 10^6, b_5 \approx 1.1 \cdot 10^6, b_7 \approx 3.7 \cdot 10^5, b_9 \approx 1.5 \cdot 10^5,$$

finally we get $a_{100} = 9999.9...$, $b_{100} = 10000.1...$, while $\mathcal{G}_{4/5}(a_0, b_0) = 10^4$.

But the answer to the above Problem can be also negative.

Example 13. We have ${}^{\mathcal{G}_{-1}}\mathcal{G}_2 = \mathcal{G}$ but taking $a_0 = 10, b_0 = 1$ and

$$a_{n+1} = \mathcal{G}_2(a_n, b_n) \text{ and } b_{n+1} = \mathcal{G}(a_n, b_n),\ n \geq 0,$$

we get $a_3 = 10^9, b_3 = 4 \cdot 10^6$, and the sequences are divergent. In this case $\mathcal{G}_2$ and $\mathcal{G}_{-1}$ are not means.

3.12 OTHER GENERALIZATIONS OF DOUBLE SEQUENCES

It is well known that, before 1799, C.F. Gauss studied the $\mathcal{AGM}$ of complex numbers. As it is shown in Cox (1984), if for two complex numbers a and b we take $a_0 = a$, $b_0 = b$ and

$$a_{n+1} = \mathcal{A}(a_n, b_n),\ b_{n+1} = \mathcal{G}(a_n, b_n),\ n \geq 0,$$

we get uncountable many sequences $(a_n)_{n\geq 0}$ and $(b_n)_{n\geq 0}$, because we have two choices for b_{n+1} for all $n \geq 0$. One of them is special. If $a, b \in \mathbb{C}^*$ and $a \neq \pm b$, a square root b_{n+1} of $a_n \cdot b_n$ is called **the right choice** if $|a_n - b_n| \leq |a_n + b_n|$ and, when $|a_n - b_n| = |a_n + b_n|$, we have $Im(b_n/a_n) > 0$. A pair of sequences $(a_n)_{n\geq 0}$ and $(b_n)_{n\geq 0}$ is called **good** if b_{n+1} is the right choice for $\sqrt{a_n b_n}$ for all but finitely many $n \geq 0$. The following result was proved in Cox (1984).

Theorem 99. *If $a, b \in \mathbb{C}^*$ and $a \neq \pm b$, then any pair of sequences $(a_n)_{n\geq 0}$ and $(b_n)_{n\geq 0}$ converge to a common limit, and this common limit is non-zero if and only if $(a_n)_{n\geq 0}$ and $(b_n)_{n\geq 0}$ are good sequences.*

Similarly, in Phillips (1984) the complex Archimedean harmonic-geometric mean is treated. For two complex numbers a and b, we take $a_0 = a$, $b_0 = b$ and

$$a_{n+1} = \mathcal{H}(a_n, b_n),\ b_{n+1} = \mathcal{G}(a_{n+1}, b_n),\ n \geq 0.$$

Excluding, as above, the trivial case $ab(a^2 - b^2) = 0$, let us denote

$$a = \lambda \cdot \tanh z,\ b = \lambda \cdot \sinh z,$$

with $z = x + iy$, $x \geq 0$ and $-\pi < y \leq \pi$. We may choose the value of the square root at each stage so that

$$a_n = 2^n \lambda \tanh \frac{z}{2^n}, \; b_n = 2^n \lambda \sinh \frac{z}{2^n},$$

thus the complex sequences $(a_n)_{n \geq 0}$ and $(b_n)_{n \geq 0}$ tend to the common limit λ. This depends again on the choice of the square root for each value of b_{n+1}, $n \geq 0$.

Using some of these ideas, in Toader (1986) means for complex numbers were defined and complex double sequences were studied. The natural general frame to develop the axiomatic study of means and its application to double sequences is that of lattices, which was considered in Toader (1987a).

Remark 97. A generalization of double sequences in another direction was given in Schlesinger (1911). Namely, a triplet of sequences is considered, defined as follows:

$$a_{n+1} = \frac{a_n + b_n + c_n}{3}, \; b_{n+1} = \left(\frac{a_n b_n + b_n c_n + c_n a_n}{3} \right)^{1/2},$$
$$c_{n+1} = (a_n b_n c_n)^{1/3}, \; n \geq 0,$$

where $a_0 = a$, $b_0 = b$ and $c_0 = c$ are three given positive numbers. It is proved that the sequences are convergent to the same limit. A similar method was used in Ory (1938) where the Heron's method was extended for higher order roots. An $\mathcal{AGH}$ mean was defined in Dmitrieva and Malozemov (1997). More generally, in Sternberg (1919), Gustin (1947), Bellman (1956–1957), Everett and Metropolis (1971), Wimp (1985), and Toader (1987a) arbitrary multiple sequences are considered.

Chapter 4

Integral means

ABSTRACT

We present a general mean that has an integral representation, which was given in Toader and Toader (2002a). This definition allows to construct means, but also pre-means. We also study some properties of integral means.

The arithmetic–geometric mean has the representation

$$\mathcal{A} \otimes \mathcal{G}(a,b) = \left[\frac{1}{2\pi}\int_0^{2\pi} \frac{d\theta}{\sqrt{a^2 \cdot \cos^2\theta + b^2 \cdot \sin^2\theta}}\right]^{-1}, \tag{4.1}$$

as we saw in the first chapter.

This expression was generalized several times (see for example Rassias and Kim, 2008) to give integral representation of many other means. In what follows we shall present a general mean of this type which was given in Toader and Toader (2002a). The definition uses three auxiliary functions. It allows to construct not only means but also pre-means. In fact, we obtain all the integral pre-means which are given in: Kim (1999), Kim and Ume (2000), Liu (2001, 2001a), Toader (2001, 2002), Kim and Rassias (2002), Rassias and Kim (2003).

4.1 THE DEFINITION OF THE INTEGRAL MEAN

Let $r : J \to J'$, $q : J' \to J''$, and $p : J' \to \mathbb{R}$ be three strictly monotonous functions. If we denote the function

$$f(a,b;p,q,r) = \frac{1}{2\pi}\int_0^{2\pi} p \circ q^{-1}[q \circ r(a) \cdot \cos^2\theta + q \circ r(b) \cdot \sin^2\theta]d\theta, \tag{4.2}$$

then

$$M_{p,q,r}(a,b) = r^{-1} \circ p^{-1}[f(a,b;p,q,r)] \tag{4.3}$$

defines a mean on J.

Let us consider some special cases. First of all, for $r = e_1$ let us denote

$$f(a,b;p,q) = f(a,b;p,q,e_1) \text{ and } M_{p,q} = M_{p,q,e_1}. \tag{4.4}$$

Means in Mathematical Analysis. http://dx.doi.org/10.1016/B978-0-12-811080-5.00004-9

If moreover $q = e_n$, we denote $M_{p,q} = M_{p,n}$. The arithmetic–geometric mean of Gauss is then obtained for $n = 2$ and $p = e_{-1}$. For $n = -2$ and $p = e_{-2}$ the mean can be found in Pólya and Szegö (1951). The case $n = 1$ and $p = \log$ was studied in Carlson (1975). The essential step was done in Haruki (1991) by considering $M_{p,n}$ for $n = 2$ with an arbitrary p. The values $n = -1$ and $n = 1$ were studied in Haruki and Rassias (1995) and Haruki and Rassias (1996). The case of an arbitrary n was studied in Toader (1998) and continued in Toader and Rassias (1999). In Toader et al. (2001) the definition of $M_{p,q}$ is given while the general construction of $M_{p,q,r}$ is given in Toader and Toader (2002a).

On the other hand, as it was remarked in Toader and Toader (2002), $M_{p,q,r}$ can be defined as an M-quasi mean. Indeed, for $q = e_1$ let us denote

$$f(a,b;p) = f(a,b;p,e_1) \text{ and } M_p = M_{p,e_1}. \tag{4.5}$$

We have

$$f(a,b;p,q,r) = f[(q \circ r)(a),(q \circ r)(b); p \circ q^{-1}]$$

and so

$$M_{p,q,r} = M_{p \circ q^{-1}}(q \circ r),$$

thus $M_{p,q,r}$ is a quasi-$M_{p \circ q^{-1}}$ mean.

4.2 A RECURRENCE FORMULA

The above results were used in Toader and Toader (2002a) for the determination of explicit values of some means $M_{p,q,r}$. For this we need to make several steps.

First of all, from Haruki (1991), Haruki and Rassias (1995), and Haruki and Rassias (1996) we can deduce the following values:

$$f(a,b;e_1) = \frac{a+b}{2},\ f(a,b;e_2) = \frac{3a^2+2ab+3b^2}{8},$$
$$f(a,b;e_{-1}) = \frac{1}{\sqrt{ab}},\ f(a,b;e_{-2}) = \frac{a+b}{2\cdot\left(\sqrt{ab}\right)^3},$$
$$f(a,b;e_o) = \log\left(\frac{\sqrt{a}+\sqrt{b}}{2}\right)^2,$$
$$f\left(a,b;e_{-1/2}\right) = \frac{2}{\pi\sqrt{a}}K\left(\sqrt{1-\frac{b}{a}}\right)$$

and

$$f\left(a,b;e_{1/2}\right)=\frac{2\sqrt{a}}{\pi}E\left(\sqrt{1-\frac{b}{a}}\right),$$

where K and E are the complete elliptic integral of the first, respectively of the second kind, given by (1.26) and (1.34).

Then, to determine $f(a,b;e_n)$ for another value of n, we use a recurrence formula. It was given in Toader (1998).

Theorem 100. *If n is a real number different from* 0, 1, *and* 2, *then*

$$f(a,b;e_n)=\left(2-\frac{1}{n}\right)\frac{a+b}{2}\cdot f(a,b;e_{n-1})-\left(1-\frac{1}{n}\right)\cdot ab\cdot f(a,b;e_{n-2}) \tag{4.6}$$

is valid for all positive numbers a,b.

Proof. We have

$$f(a,b;e_n)=\frac{1}{2\pi}\int_0^{2\pi}(a\cdot\cos^2\theta+b\cdot\sin^2\theta)^n d\theta$$
$$=\frac{1}{2\pi}\left(\frac{a+b}{2}\right)^n\int_0^{2\pi}\left(1+\frac{a-b}{a+b}\cdot\cos 2\theta\right)^n d\theta.$$

Denoting

$$k=\frac{a-b}{a+b}\text{ and }J_n=\int_0^{2\pi}(1+k\cdot\cos 2\theta)^n\,d\theta,$$

we get

$$f(a,b;e_n)=\frac{1}{2\pi}\cdot\left(\frac{a+b}{2}\right)^n\cdot J_n.$$

For J_n we have

$$J_n=\int_0^{2\pi}(1+k\cdot\cos 2\theta)\cdot(1+k\cdot\cos 2\theta)^{n-1}\,d\theta$$
$$=J_{n-1}+\frac{k}{2}\cdot\int_0^{2\pi}(\sin 2\theta)'\cdot(1+k\cdot\cos 2\theta)^{n-1}\,d\theta$$
$$=J_{n-1}+\frac{k}{2}\cdot\left[\sin 2\theta\cdot(1+k\cdot\cos 2\theta)^{n-1}\right]\Big|_0^{2\pi}$$
$$+k^2(n-1)\int_0^{2\pi}(\sin 2\theta)^2\cdot(1+k\cdot\cos 2\theta)^{n-2}\,d\theta=J_{n-1}$$

$$+k^2(n-1)\cdot J_{n-2}-(n-1)\int_0^{2\pi}(k\cdot\cos 2\theta)^2\cdot(1+k\cdot\cos 2\theta)^{n-2}\,d\theta.$$

So,

$$n\cdot J_n=(2n-1)\cdot J_{n-1}+(n-1)(k^2-1)\cdot J_{n-2},$$

which gives (4.6). □

Remark 98. Using the recurrence formula and the expressions given above, we can determine, step by step, the value of $f(a,b;e_{n/2})$ for all $n\in\mathbb{Z}$. Moreover, if $p\circ q^{-1}=e_{n/2}$, we have

$$f(a,b;e_{n/2}\circ q,q,r)=f(q\circ r(a),q\circ r(b);e_{n/2}).$$

So we can use the above values to get the expression of any term of this type and the value of the corresponding means. For example, we have

$$M_{e_{-3}\circ q,q,r}(a,b)$$

$$=r^{-1}\circ q^{-1}\left(\frac{2[(q\circ r)(a)\cdot(q\circ r)(b)]^{5/6}}{[3((q\circ r)(b))^2+2\cdot(q\circ r)(a)\cdot(q\circ r)(b)+3((q\circ r)(a))^2]^{1/3}}\right).$$

Remark 99. We cannot use the recurrence formula for the determination of $M_{p,n}$ for $n=0$. In Toader and Rassias (1999) a connection between these means and the Bessel functions is given. We have $M_{e_0,e_0}=\mathcal{G}$ while, for $m\neq 0$,

$$M_{e_m,e_0}(a,b)=\left(\frac{1}{2\pi}\int_0^{2\pi}a^{m\cos^2\theta}b^{m\sin^2\theta}d\theta\right)^{1/m}$$

$$=\mathcal{G}(a,b)\left[\frac{1}{\pi}\int_0^{\pi}\left(\frac{a}{b}\right)^{(m/2)\cos\varphi}d\varphi\right]^{1/m}=\mathcal{G}(a,b)\left(\frac{1}{\pi}\int_0^{\pi}e^{X\cos\varphi}d\varphi\right)^{1/q}$$

where $X=(m/2)\ln(a/b)$. From Gradshteyn and Ryzhik (1994) we deduce that

$$\frac{1}{\pi}\int_0^{\pi}e^{X\cos\varphi}d\varphi=J_0(-iX),$$

where J_0 is the Bessel function of the first kind. Thus,

$$M_{e_m,e_0}(a,b)=\mathcal{G}(a,b)\,[J_0(-iX)]^{1/m}=\mathcal{G}(a,b)\left[\sum_{k=0}^{\infty}\left(\frac{X}{2}\right)^{2k}\frac{1}{(k!)^2}\right]^{1/m}$$

$$=\sqrt{ab}\left[\sum_{k=0}^{\infty}\left(\frac{m}{4}\ln\frac{a}{b}\right)^{2k}\frac{1}{(k!)^2}\right]^{1/m}.$$

4.3 GAUSS' FUNCTIONAL EQUATION

In the study of the arithmetic–geometric mean we have distinguished three main mathematical objects: the mean, its integral representation, and the Gauss' functional equation (which makes the connection between the first two). We have presented generalizations of the integral representation formula. The Gauss' functional equation was also generalized several times. We give here the most general functional equation

$$F(M(a,b), N(a,b)) = F(a,b) \tag{4.7}$$

as it was considered in Toader and Toader (2002a). Assuming that M and N are two arbitrary pre-means, the following results were proved.

Lemma 19. *If the function F is two times differentiable and verifies the functional equation (4.7), then the function F is a solution of the differential equation*

$$\begin{gathered}F_{aa}(c,c)\cdot M_a(c,c)\cdot M_b(c,c)+F_{ab}(c,c)\cdot[M_a(c,c)\cdot N_b(c,c)\\+M_b(c,c)\cdot N_a(c,c)-1]+F_{bb}(c,c)\cdot N_a(c,c)\cdot N_b(c,c)+F_a(c,c)\cdot M_{ab}(c,c)\\+F_b(c,c)\cdot N_{ab}(c,c)=0.\end{gathered}$$

Proof. Taking in (4.7) the partial derivatives in respect to a we obtain

$$F_a[M(a,b),N(a,b)]\cdot M_a(a,b)+F_b[M(a,b),N(a,b)]\cdot N_a(a,b)=F_a(a,b).$$

Taking again the derivatives with respect to b it follows that

$$\begin{gathered}\{F_{aa}[M(a,b),N(a,b)]\cdot M_b(a,b)\\+F_{ab}[M(a,b),N(a,b)]\cdot N_b(a,b)\}\cdot M_a(a,b)\\+\{F_{ab}[M(a,b),N(a,b)]\cdot M_b(a,b)\\+F_{bb}[M(a,b),N(a,b)]\cdot N_b(a,b)\}\cdot N_a(a,b)\\+F_a[M(a,b),N(a,b)]\cdot M_{ab}(a,b)\\+F_b[M(a,b),N(a,b)]\cdot N_{ab}(a,b)=F_{ab}(a,b).\end{gathered}$$

For $a=b=c$ we obtain the desired formula. □

Corollary 69. *If the function F is two times differentiable and verifies the functional equation (4.7) for symmetric pre-means M and N, then the function F is a solution of the differential equation*

$$\begin{gathered}F_{aa}(c,c)-2F_{ab}(c,c)+F_{bb}(c,c)+4F_a(c,c)\cdot M_{ab}(c,c)\\+4F_b(c,c)\cdot N_{ab}(c,c)=0.\end{gathered}$$

Proof. As we saw in (2.27), if M is a symmetric pre-mean then

$$M_a(c,c) = M_b(c,c) = 1/2. \qquad \square$$

In what follows we assume that the function p is two times differentiable. As in Haruki (1991), we can prove the following result.

Lemma 20. *The function f defined by (4.5) has the following partial derivatives:*

$$f_a(c,c;p) = f_b(c,c;p) = \frac{1}{2} \cdot p'(c),$$
$$f_{aa}(c,c;p) = f_{bb}(c,c;p) = \frac{3}{8} \cdot p''(c)$$

and

$$f_{ab}(c,c;p) = \frac{1}{8} \cdot p''(c).$$

Proof. From the definition, we have

$$f(a,b;p) = \frac{1}{2\pi} \int_0^{2\pi} p(a \cdot \cos^2\theta + b \cdot \sin^2\theta)\, d\theta,$$

thus

$$f_a(a,b;p) = \frac{1}{2\pi} \int_0^{2\pi} p'(a \cdot \cos^2\theta + b \cdot \sin^2\theta) \cdot \cos^2\theta\, d\theta,$$
$$f_{aa}(a,b;p) = \frac{1}{2\pi} \int_0^{2\pi} p''(a \cdot \cos^2\theta + b \cdot \sin^2\theta) \cdot \cos^4\theta\, d\theta$$

and

$$f_{ab}(a,b;p) = \frac{1}{2\pi} \int_0^{2\pi} p''(a \cdot \cos^2\theta + b \cdot \sin^2\theta) \cdot \cos^2\theta \cdot \sin^2\theta\, d\theta.$$

Setting $a = b = c$, we get the desired relations. $\square$

Using them, the following results were proved in Toader and Toader (2002a).

Lemma 21. *The function f defined by (4.2) has the following partial derivatives:*

$$f_a(c,c;p,q,r) = f_b(c,c;p,q,r) = \frac{1}{2} \cdot (p \circ r)'(c),$$
$$f_{aa}(c,c;p,q,r) = f_{bb}(c,c;p,q,r) = \frac{3}{8} \cdot (p \circ r)''(c)$$

$$+\frac{1}{8}\cdot(p\circ r)'(c)\cdot\frac{(q\circ r)''(c)}{(q\circ r)'(c)}$$

and

$$f_{ab}(c,c;p,q,r)=\frac{1}{8}\cdot(p\circ r)''(c)-\frac{1}{8}\cdot(p\circ r)'(c)\cdot\frac{(q\circ r)''(c)}{(q\circ r)'(c)}. \tag{4.8}$$

Proof. If we consider

$$g(a,b)=h(u(a),v(b)),$$

we have the following well-known formulas:

$$g_a(a,b)=h_u(u(a),v(b))\cdot u'(a),$$
$$g_{aa}(a,b)=h_{uu}(u(a),v(b))\cdot u'^2(a)+h_u(u(a),v(b))\cdot u''(a)$$

and

$$g_{ab}(a,b)=h_{uv}(u(a),v(b))\cdot u'(a)\cdot v'(b).$$

Using them for (4.2), we get:

$$f_a(c,c;p,q,r)=\frac{1}{2}\cdot(p\circ q^{-1})'[(q\circ r)(c)]\cdot(q\circ r)'(c)$$
$$=\frac{1}{2}\cdot[(p\circ q^{-1})\circ(q\circ r)]'(c)=\frac{1}{2}\cdot(p\circ r)'(c),$$
$$f_{aa}(c,c;p,q,r)=\frac{3}{8}\cdot(p\circ q^{-1})''[(q\circ r)(c)]\cdot(q\circ r)'^2(c)$$
$$+\frac{1}{2}\cdot(p\circ q^{-1})'[(q\circ r)(c)]\cdot(q\circ r)''(c)$$
$$=\frac{3}{8}\cdot\{(p\circ q^{-1})''[(q\circ r)(c)]\cdot(q\circ r)'^2(c)+(p\circ q^{-1})'[(q\circ r)(c)]$$
$$\cdot(q\circ r)''(c)\}+\frac{1}{8}\cdot(p\circ q^{-1})'[(q\circ r)(c)]\cdot(q\circ r)''(c)$$
$$=\frac{3}{8}\cdot[(p\circ q^{-1})\circ(q\circ r)]''(c)+\frac{1}{8}\cdot[(p\circ q^{-1})\circ(q\circ r)]'(c)$$
$$\cdot\frac{(q\circ r)''(c)}{(q\circ r)'(c)}=\frac{3}{8}\cdot(p\circ r)''(c)+\frac{1}{8}\cdot(p\circ r)'(c)\cdot\frac{(q\circ r)''(c)}{(q\circ r)'(c)}$$

and

$$f_{ab}(c,c;p,q,r)=\frac{1}{8}\cdot(p\circ q^{-1})''[(q\circ r)(c)]\cdot(q\circ r)'^2(c)$$
$$=\frac{1}{8}\cdot\{[(p\circ q^{-1})\circ(q\circ r)]''(c)-(p\circ q^{-1})'[(q\circ r)(c)]\cdot(q\circ r)''(c)\}$$

$$= \frac{1}{8} \cdot (p \circ r)''(c) - \frac{1}{8} \cdot (p \circ r)'(c) \cdot \frac{(q \circ r)''(c)}{(q \circ r)'(c)}. \qquad \square$$

Combining the previous results, we obtain the following result.

Theorem 101. *If the function f, defined by (4.2), verifies the functional equation (4.7), then the following relation*

$$\begin{aligned}
&(p \circ r)''(c) \cdot [3 \cdot M_a(c,c) \cdot M_b(c,c) + M_a(c,c) \cdot N_b(c,c) \\
&+ M_b(c,c) \cdot N_a(c,c) + 3 \cdot N_a(c,c) \cdot N_b(c,c) - 1] + (p \circ r)'(c) \\
&\cdot \{4 \cdot [M_{ab}(c,c) + N_{ab}(c,c)] + \frac{(q \circ r)''(c)}{(q \circ r)'(c)} \cdot [M_a(c,c) \cdot M_b(c,c) \\
&+ N_a(c,c) \cdot N_b(c,c) - M_a(c,c) \cdot N_b(c,c) - M_b(c,c) \cdot N_a(c,c) + 1]\} = 0
\end{aligned}$$

holds.

Corollary 70. *If the function f, defined by (4.2), verifies the functional equation (4.7), where the pre-means M and N are symmetric, then*

$$(p \circ r)''(c) + (p \circ r)'(c) \cdot \left\{ \frac{(q \circ r)''(c)}{(q \circ r)'(c)} + 4 \cdot [M_{ab}(c,c) + N_{ab}(c,c)] \right\} = 0.$$

Remark 100. Most of the usual means M have the property (2.30), that is, there exists a constant $\alpha = \alpha(M)$ such that

$$M_{ab}(c,c) = -\frac{\alpha(M)}{c}.$$

If M and N are such means that $q = e_n$ and $r = e_m$, with $nm \neq 0$, the above relation becomes

$$(p \circ r)''(c) + (p \circ r)'(c) \cdot \frac{n - 1 - 4 \cdot [\alpha(M) + \alpha(N)]}{c} = 0,$$

which gives the following:

Corollary 71. *If the function f defined by (4.2) verifies the functional equation (4.7), where $q = e_n$ and $r = e_m$, with $nm \neq 0$, and the symmetric pre-means M and N are such that*

$$M_{ab}(c,c) = -\frac{\alpha(M)}{c}, \quad N_{ab}(c,c) = -\frac{\alpha(N)}{c},$$

then

$$p(c) = k \cdot c^{\{1+4\cdot[\alpha(M)+\alpha(N)]\}/m-1} + h,$$

where k and h are arbitrary constants.

Remark 101. These results are generalizations of all similar properties proved in the papers: Haruki (1991), Haruki and Rassias (1995), Haruki and Rassias (1996), Toader (1998), Toader and Rassias (1999), Kim (1999), Kim and Ume (2000), Ume and Kim (2000), Toader et al. (2001), or Toader (2002). In all of them, for special functions q and r and special means M and N, the function p is determined, as a solution of the above differential equation. In some cases it is proved that we really get a solution of the given functional equation. For example, the following consequence holds.

Corollary 72. *If the function f given by (4.4) is a solution of the functional equation*

$$F(\mathcal{P}_t(a,b), \mathcal{P}_s(a,b)) = F(a,b), \tag{4.9}$$

p has a continuous second order derivative, and $q = e_n$, then

$$p = \alpha\, e_{t+s-n} + \beta, \tag{4.10}$$

with α, β arbitrary real constants.

The condition is also sufficient in some cases as it was proved in Toader and Rassias (1999).

Theorem 102. *The function f given by (4.4) with p given by (4.10) and $q = e_n$, verifies the relation (4.9) if $n \neq 0$, $t = n$, and $s = -n$.*

Proof. It is enough to set $p = e_{-n}$. Then

$$\begin{aligned}
&2f(a,b;e_{-n},e_n)\\
&= \frac{1}{2\pi}\left[\int_0^{2\pi} \frac{d\theta}{a^n \sin^2\theta + b^n \cos^2\theta} + \int_0^{2\pi} \frac{d\theta}{b^n \sin^2\theta + a^n \cos^2\theta}\right]\\
&= \frac{1}{2\pi}\int_0^{2\pi} \frac{(a^n + b^n)d\theta}{a^n b^n(\sin^4\theta + \cos^4\theta) + (a^{2n} + b^{2n})\sin^2\theta\cos^2\theta}\\
&= \frac{1}{\pi}\int_0^{2\pi} \frac{d\theta}{[\mathcal{P}_{-n}(a,b)]^n \cos^2 2\theta + [\mathcal{P}_n(a,b)]^n \sin^2 2\theta}.
\end{aligned}$$

If we set $\varphi = 2\theta$, we obtain

$$\begin{aligned}
f(a,b;e_{-n},e_n) &= \frac{1}{4\pi}\int_0^{4\pi} \frac{d\varphi}{[\mathcal{P}_{-n}(a,b)]^n \cos^2\varphi + [\mathcal{P}_n(a,b)]^n \sin^2\varphi}\\
&= f(\mathcal{P}_{-n}(a,b), \mathcal{P}_n(a,b); e_{-n}, e_n).
\end{aligned}$$

Therefore (4.9) is verified in this case. □

Remark 102. The result was proved for $n = 1$, $t = 1$, and $s = -1$ in Haruki and Rassias (1995).

Remark 103. A relation of another type than (4.7) was proved in Carlson (1975). It is proved that $f(a, b; \log, e_1)$ verifies

$$F(\mathcal{A}^2(a,b), \mathcal{G}^2(a,b)) = 2F(a,b).$$

4.4 SPECIAL INTEGRAL MEANS

Another problem studied in all the papers related to integral means of (4.1) type is that of the determination of functions p, q and r for which the mean $M_{p,q,r}$ reduces to a given known mean, for example a power mean $\mathcal{P}_m$. Using some results from the previous section we can prove the following general results.

Theorem 103. *If for a given mean N we have*

$$M_{p,q,r} = N$$

then the functions p, q, and r verify the relation

$$\begin{aligned}(p \circ r)''(c) \cdot (q \circ r)'(c) \cdot [1 - 8 \cdot N_a(c,c) \cdot N_b(c,c)] \\ = (p \circ r)'(c) \cdot \left[(q \circ r)''(c) + 8 \cdot N_{ab}(c,c) \cdot (q \circ r)'(c)\right].\end{aligned} \tag{4.11}$$

Proof. We have

$$f(a,b; p, q.r) = p\left[r(N(a,b))\right].$$

Taking the partial derivative with respect to a, we have

$$f_a(a,b; p, q.r) = p'\left[r(N(a,b))\right] \cdot r'(N(a,b)) \cdot N_a(a,b).$$

If we take now the partial derivative with respect to b, we get

$$\begin{aligned}f_{ab}(a,b; p, q.r) = p''\left[r(N(a,b))\right] \cdot \left[r'(N(a,b))\right]^2 \cdot N_a(a,b) \\ \cdot N_b(a,b) + p'\left[r(N(a,b))\right] \cdot r''(N(a,b)) \cdot N_a(a,b) \cdot N_b(a,b) \\ + p'\left[r(N(a,b))\right] \cdot r'(N(a,b)) \cdot N_{ab}(a,b).\end{aligned}$$

Setting $a = b = c$ we have

$$\begin{aligned}f_{ab}(c,c; p, q.r) = p''\left[r(c)\right] \cdot \left[r'(c)\right]^2 \cdot N_a(c,c) \cdot N_b(c,c) \\ + p'\left[r(c)\right] \cdot r''(c) \cdot N_a(c,c) \cdot N_b(c,c) + p'\left[r(c)\right] \cdot r'(c) \cdot N_{ab}(c,c),\end{aligned}$$

thus

$$f_{ab}(c,c; p, q.r) = (p \circ r)''(c) \cdot N_a(c,c) \cdot N_b(c,c) + (p \circ r)'(c) \cdot N_{ab}(c,c).$$

Using (4.8) we obtain the desired relation. □

Corollary 73. *If for a symmetric mean N we have*

$$M_{p,q,r} = N$$

then the functions p, q, and r verify the relation

$$\begin{aligned}(p \circ r)''(c) \cdot (q \circ r)'(c) + (p \circ r)'(c) \\ \cdot \left[(q \circ r)''(c) + 8 \cdot N_{ab}(c,c) \cdot (q \circ r)'(c)\right] = 0\end{aligned} \quad (4.12)$$

Proof. As we saw in (2.27), if N is a symmetric pre-mean, then

$$N_a(c,c) = N_b(c,c) = 1/2$$

so that (4.11) reduces to (4.12). □

Corollary 74. *If the symmetric mean N is such that*

$$N_{ab}(c,c) = -\frac{\alpha(N)}{c}$$

and

$$M_{p,q,r} = N$$

where

$$q \circ r = e_n,$$

then

$$p \circ r = k \cdot e_{2-n+8\cdot\alpha(N)} + h, \; k, h \in \mathbb{R}. \quad (4.13)$$

Proof. The new hypotheses simplify (4.12) to

$$c \cdot (p \circ r)''(c) + (p \circ r)'(c) \cdot [n - 1 - 8 \cdot \alpha(N)] = 0$$

which gives (4.13). □

Corollary 75. *If the mean $M_{p,n}$ reduces at the power mean $\mathcal{P}_m$ for arbitrary n, then*

$$p = \alpha\, e_{2m-n} + \beta, \; \alpha, \beta \in \mathbb{R}.$$

Proof. We know that

$$\alpha(\mathcal{P}_m) = \frac{m-1}{4}$$

and we apply (4.13). □

Remark 104. In some cases the above conditions are also sufficient to obtain the mean $\mathcal{P}_m$. This was proved in Haruki (1991) for $n = 2$ and $m = 0, 1$, or 2 and in Haruki and Rassias (1996) for $n = 1$ and $m = 0, 1$, or $1/2$ and for $n = -1$ and $m = 0, -1$, or $-1/2$. They are special cases of the following result proved in Toader and Rassias (1999).

Theorem 104. *The mean $M_{p,n}$ reduces to the power mean $\mathcal{P}_m$ for arbitrary n if*

$$p = \alpha\, e_{2m-n} + \beta,\ \alpha, \beta \in \mathbb{R}$$

and m takes one of the following values: i) $m = 0$; ii) $m = n$; or iii) $m = n/2$.

Proof. It is enough to take $p = e_{2m-n}$. i) We have

$$M_{e_{-n},n}(a,b) = \left[f(a,b;e_{-n},n)\right]^{-1/n}.$$

As it was remarked in Toader (1998),

$$f(a,b;e_m,e_n) = f(a^n,b^n;e_{m/n}), \tag{4.14}$$

thus

$$M_{e_{-n},n}(a,b) = \left[f(a^n,b^n;e_{-1})\right]^{-1/n}.$$

The relation

$$f(a,b;e_{-1}) = \frac{1}{\sqrt{ab}},$$

which was given before, implies

$$M_{e_{-n},n}(a,b) = \left[\frac{1}{\sqrt{a^n b^n}}\right]^{-1/n} = \sqrt{ab} = \mathcal{P}_o(a,b).$$

ii) Similarly, using again (4.14), we have

$$M_{e_n,n}(a,b) = \left[f(a,b;e_n,e_n)\right]^{1/n} = \left[f(a^n,b^n;e_1)\right]^{1/n} = \mathcal{P}_n(a,b).$$

iii) As $e_0 = \log$, we obtain

$$M_{e_o,n}(a,b) = \exp\left[f(a,b;e_o,n)\right].$$

Now, instead of (4.14) we have the relation

$$f(a,b;e_o,e_n) = \frac{1}{n} f(a^n,b^n;e_o)$$

and using the formula

$$f(a,b;e_o)=\log\left(\frac{a^{1/2}+b^{1/2}}{2}\right)^2,$$

it follows that

$$M_{e_o,n}(a,b)=\exp\left[\frac{1}{n}\log\left(\frac{a^{n/2}+b^{n/2}}{2}\right)^2\right]=\mathcal{P}_{n/2}(a,b).\qquad\square$$

Remark 105. It is an open question if the property is valid also for other values of m. For example, it is known (see Haruki and Rassias, 1996) that $M_{e_{-1},2}\neq\mathcal{P}_{1/2}$ and $M_{e_{-4},2}\neq\mathcal{P}_{-1}$.

4.5 COMPARISON OF INTEGRAL MEANS

The means $M_{p,q}$ are special cases of the means $M_{p,\mathcal{R}}$ defined in Toader (1998a) by (2.14). Indeed, let us consider a strictly monotonous function q and take the family of means $\mathfrak{R}_q=\{R_{q,\theta}:\theta\in[0,2\pi]\}$, defined by

$$R_{q,\theta}(a,b)=q^{-1}[q(a)\cdot\cos^2\theta+q(b)\cdot\sin^2\theta].$$

The mean $M_{p,\mathfrak{R}}$ for $\mathfrak{R}=\mathfrak{R}_q$ is just $M_{p,q}$.

We want to give some applications of the inequalities (2.16) and (2.17), for this special case.

Corollary 76. *If $k<h$ then*

$$M_{e_k,q}<M_{e_h,q}\tag{4.15}$$

for every strictly monotonous function q.

Proof. If $0<k<h$, then $e_h\circ e_k^{-1}=e_{h/k}$ is convex and $e_h^{-1}=e_{1/h}$ is increasing. For $k=0<h$, $(e_h\circ e_o^{-1})(x)=\exp(hx)$ is also convex and $e_h^{-1}=e_{1/h}$ increasing. If $k<0=h$, again $e_o\circ e_k^{-1}=k^{-1}e_o$ is convex and $e_o^{-1}=\exp$ is increasing. Finally if $k<h<0$, $e_h\circ e_k^{-1}=e_{h/k}$ is concave but $e_h^{-1}=e_{1/h}$ is decreasing. In all the cases we get (4.15). $\square$

Lemma 22. *The means $R_{e_n,\theta}$ are increasing with n for each $\theta\in[0,2\pi]$.*

Proof. We have

$$R_{e_n,\theta}(a,b)=\begin{cases}a\left[\cos^2\theta+\left(\frac{b}{a}\right)^n\sin^2\theta\right]^{1/n}, & n\neq0\\ a\cdot\left(\frac{b}{a}\right)^{\sin^2\theta}, & n=0\end{cases}$$

thus $R_{e_n,\theta}(a,b) = a \cdot f(n)$, where

$$f(n) = \mathcal{P}_{n,t}(c,1)$$

with $c = b/a$ and $t = \sin^2\theta$. The function f is increasing. □

Corollary 77. *If $n < m$ we have $M_{p,e_n} < M_{p,e_m}$ for every monotonous function p.*

Other inequalities with integral means can be found for example in Toader and Raşa (1996).

4.6 INTEGRAL PRE-MEANS

In Kim (1999), the expression considered was

$$\mathcal{K}_{p,n}(a,b) = \frac{1}{\mathcal{H}(a,b)} \cdot p^{-1}(f(a,b;p,e_{2n})) \tag{4.16}$$

for $n = \pm 1$. It can be written also as

$$\mathcal{K}_{p,n} = \frac{M^2_{p,e_n,e_2}}{\mathcal{H}}$$

and, as it was proved in Toader (2001), generally this is not a mean, it is only a pre-mean.

In what follows we present a general definition of pre-means of this type given in Toader and Toader (2002a). It contains as special cases all the other generalizations of (4.16) of which we are aware.

Let the functions p, q, and r be as in (4.2) and N and R be two arbitrary pre-means. Let us define the expression

$$M(p,q,r;N,R) = R \cdot \frac{p^{-1}(f(a,b;p,q,r))}{r(N)}.$$

As

$$M(p,q,r;N,R) = R \cdot \frac{r\left(M_{p,q,r}\right)}{r(N)} = R \cdot \frac{r\left(M_{p\circ q^{-1}}(q \circ r)\right)}{r(N)},$$

it is easy to check that it represents a pre-mean.

Remark 106. If we take $r = e_1$, we get simply

$$M(p,q,e_1;N,R) = R \cdot \frac{p^{-1}(f(a,b;p,q))}{N} = R \cdot \frac{M_{p,q}}{N},$$

which are operations only with means. But for (4.16) we have

$$\mathcal{K}_{p,n} = M(p, e_n, e_2; \mathcal{H}, \mathcal{H}).$$

For arbitrary n this was studied in Kim and Ume (2000). In Toader (2002), $\mathcal{H}$ was replaced by an arbitrary mean N. In Ume and Kim (2000), the case $M(p, e_2, e_{\pm 1}; \mathcal{G}, \mathcal{A}_t)$ was studied. In Liu (2001), also the cases $M(p, e_1, e_2; \mathcal{H}, \mathcal{H})$ and $M(p, e_{-1}, e_2; \mathcal{H}, \mathcal{H})$ were studied, while in Kim and Rassias (2002), the case $M(p, e_n, e_k; \mathcal{G}, \mathcal{G})$ was studied.

Remark 107. Of course, sometimes $M(p, q, r; N, R)$ is a mean. For instance, in Kim and Ume (2000), $M(p, e_{-1}, e_n; \mathcal{G}, \mathcal{G})$ is considered, that is, $\mathcal{G}^2 / M_{p,e_{-1},e_n}$ which is always a mean, namely it is the inverse mean of M_{p,e_{-1},e_n}.

Remark 108. Usually, the authors give the values of some special cases of the obtained pre-means. Using them and the recurrence formula given before, we can then deduce more such values. Indeed, we have

$$M(p, q, r; N, R)(a, b) = \frac{R(a, b)}{r(N(a, b))} \cdot p^{-1}\left(f(q \circ r(a), q \circ r(b); p \circ q^{-1})\right).$$

If $p \circ q^{-1} = e_{n/2}$, we get

$$M(e_{n/2} \circ r, q, r; N, R)(a, b)$$
$$= \frac{R(a, b)}{r(N(a, b))} \cdot q^{-1}\left[(f(q \circ r(a), q \circ rq(b); e_{n/2}))^{2/n}\right].$$

Knowing the value of $f(q \circ r(a), q \circ rq(b); e_{n/2})$, we can calculate the value of the pre-mean.

4.7 SPECIAL PRE-MEANS

A problem which is also studied for the integral pre-means defined in the above mentioned papers is that of the determination of the cases in which the integral mean reduces to a given mean. The following general result was proved in Toader and Toader (2002a).

Theorem 105. *If for given pre-means N, R, and S we have*

$$M(p, q, r; N, R) = S,$$

then the following differential relation is verified by the functions p, q, and r:

$$(p \circ r)''(c) \cdot \left[\frac{1}{8} - N'_a(c, c) \cdot N'_b(c, c)\right]$$

$$-(p\circ r)'(c)\cdot\left[\frac{1}{8}\cdot\frac{(q\circ r)''(c)}{(q\circ r)'(c)}+\frac{S'_a(c,c)-R'_a(c,c)}{c}\cdot N'_b(c,c)\right.$$
$$\left.+\frac{S'_b(c,c)-R'_b(c,c)}{c}\cdot N'_a(c,c)+N''_{ab}(c,c)\right]$$
$$=p''(r(c))\cdot\left\{\frac{r^2(c)}{c^2}\cdot\left[S'_a(c,c)-R'_a(c,c)\right]\cdot\left[S'_b(c,c)-R'_b(c,c)\right]\right.$$
$$+\frac{r(c)\cdot r'(c)}{c}\cdot[\left(S'_a(c,c)-R'_a(c,c)\right)\cdot N'_b(c,c)$$
$$\left.+(S'_b(c,c)-R'_b(c,c))\cdot N'_a(c,c)]\right\}$$
$$+p'(r(c))\cdot r(c)\cdot\left[\frac{S''_{ab}(c,c)-R''_{ab}(c,c)}{c}\right.$$
$$\left.+\frac{S'_a(c,c)R'_b(c,c)-R'_a(c,c)S'_b(c,c)-2R'_b(c,c)\left[S'_a(c,c)-R'_a(c,c)\right]}{c^2}\right].$$

Proof. In the given hypotheses we have

$$f(a,b;p,q,r)=p\left(\frac{S(a,b)}{R(a,b)}\cdot r[N(a,b)]\right).$$

Taking the partial derivative with respect to a, we have

$$f'_a(a,b;p,q,r)=p'\left(\frac{S(a,b)}{R(a,b)}\cdot r[N(a,b)]\right)$$
$$\cdot\left[\frac{S'_a(a,b)\cdot R(a,b)-R'_a(a,b)\cdot S(a,b)}{[R(a,b)]^2}\cdot r[N(a,b)]\right.$$
$$\left.+\frac{S(a,b)}{R(a,b)}\cdot r'[N(a,b)]\cdot N'_a(a,b)\right].$$

If we take then the derivative with respect to b, we obtain

$$f''_{ab}(a,b;p,q,r)=p''\left(\frac{S(a,b)}{R(a,b)}\cdot r[N(a,b)]\right)$$
$$\cdot\left[\frac{S'_a(a,b)\cdot R(a,b)-R'_a(a,b)\cdot S(a,b)}{[R(a,b)]^2}\cdot r[N(a,b)]\right.$$
$$\left.+\frac{S(a,b)}{R(a,b)}\cdot r'[N(a,b)]\cdot N'_a(a,b)\right]$$
$$\cdot\left[\frac{S'_b(a,b)\cdot R(a,b)-R'_b(a,b)\cdot S(a,b)}{[R(a,b)]^2}\cdot r[N(a,b)]\right.$$
$$\left.+\frac{S(a,b)}{R(a,b)}\cdot r'[N(a,b)]\cdot N'_b(a,b)\right]$$

$$+p'\left(\frac{S(a,b)}{R(a,b)}\cdot r[N(a,b)]\right)$$

$$\cdot\left\{\left[\frac{S''_{ab}(a,b)R(a,b)+S'_a(a,b)R'_b(a,b)-R''_{ab}(a,b)S(a,b)-R'_a(a,b)S'_b(a,b)}{[R(a,b)]^2}\right.\right.$$

$$\left.-\frac{2R'_b(a,b)\left[S'_a(a,b)\cdot R(a,b)-R'_a(a,b)\cdot S(a,b)\right]}{[R(a,b)]^3}\right]\cdot r[N(a,b)]$$

$$+\frac{S'_a(a,b)\cdot R(a,b)-R'_a(a,b)\cdot S(a,b)}{[R(a,b)]^2}\cdot r'[N(a,b)]\cdot N'_b(a,b)$$

$$+\frac{S'_b(a,b)\cdot R(a,b)-R'_b(a,b)\cdot S(a,b)}{[R(a,b)]^2}\cdot r'[N(a,b)]\cdot N'_a(a,b)$$

$$\left.+\frac{S(a,b)}{R(a,b)}\cdot\left[r''[N(a,b)]\cdot N'_a(a,b)\cdot N'_b(a,b)+r'[N(a,b)]\cdot N''_{ab}(a,b)\right]\right\}.$$

For $a=b=c$, we get

$$f''_{ab}(c,c;p,q,r)=p''(r(c))\cdot\left[\frac{S'_a(c,c)-R'_a(c,c)}{c}\cdot r(c)+r'(c)\cdot N'_a(c,c)\right]$$

$$\cdot\left[\frac{S'_b(c,c)-R'_b(c,c)}{c}\cdot r(c)+r'(c)\cdot N'_b(c,c)\right]$$

$$+p'(r(c))\left\{\left[\frac{c\cdot S''_{ab}(c,c)+S'_a(c,c)R'_b(c,c)-c\cdot R''_{ab}(c,c)-R'_a(c,c)S'_b(c,c)}{c^2}\right.\right.$$

$$\left.-\frac{2R'_b(c,c)\left[S'_a(c,c)-R'_a(c,c)\right]}{c^2}\right]\cdot r(c)+\frac{S'_a(c,c)-R'_a(c,c)}{c}\cdot r'(c)\cdot N'_b(c,c)$$

$$+\frac{S'_b(c,c)-R'_b(c,c)}{c}\cdot r'(c)\cdot N'_a(c,c)$$

$$\left.+r''(c)\cdot N'_a(c,c)\cdot N'_b(c,c)+r'(c)\cdot N''_{ab}(c,c)\right\}.$$

Replacing the first member by its known expression, we get

$$\frac{1}{8}\cdot(p\circ r)''(c)-\frac{1}{8}\cdot(p\circ r)'(c)\cdot\frac{(q\circ r)''(c)}{(q\circ r)'(c)}$$

$$=p''(r(c))\cdot\left[\frac{S'_a(c,c)-R'_a(c,c)}{c}\cdot r(c)+r'(c)\cdot N'_a(c,c)\right]$$

$$\cdot\left[\frac{S'_b(c,c)-R'_b(c,c)}{c}\cdot r(c)+r'(c)\cdot N'_b(c,c)\right]$$

$$+p'(r(c))\left\{\left[\frac{c\cdot S''_{ab}(c,c)+S'_a(c,c)R'_b(c,c)-c\cdot R''_{ab}(c,c)-R'_a(c,c)S'_b(c,c)}{c^2}\right.\right.$$

$$\left.-\frac{2R'_b(c,c)\left[S'_a(c,c)-R'_a(c,c)\right]}{c^2}\right]\cdot r(c)+\frac{S'_a(c,c)-R'_a(c,c)}{c}\cdot q'(c)\cdot N'_b(c,c)$$

$$+\frac{S'_b(c,c)-R'_b(c,c)}{c}\cdot q'(c)\cdot N'_a(c,c)$$
$$\left.+q''(c)\cdot N'_a(c,c)\cdot N'_b(c,c)+q'(c)\cdot N''_{ab}(c,c)\right\}.$$

As

$$p''(r(c))\cdot\left[r'(c)\right]^2+p'(r(c))\cdot r''(c)=(p\circ r)''(c)$$

and

$$p'(r(c))\cdot r'(c)=(p\circ r)'(c),$$

we obtain the desired result. □

Corollary 78. *If for given symmetric pre-means N, R, and S, we have*

$$M(p,q,r;N,R)=S,$$

then the following differential relation is verified by the functions p,q, and r:

$$(p\circ r)''(c)+(p\circ r)'(c)\left[\frac{(q\circ r)''(c)}{(q\circ r)'(c)}+\frac{8r(c)}{r'(c)}\cdot\frac{S''_{ab}(c,c)-R''_{ab}(c,c)}{c}+8N''_{ab}(c,c)\right]=0.$$

Proof. As the pre-means are symmetric, we use (2.27). □

Corollary 79. *If for given symmetric pre-means N, R, and S, we have*

$$M(p,e_n,e_m;N,R)=S,$$

with $nm\neq 0$ and

$$N''_{ab}(c,c)=-\frac{\alpha(M)}{c},\ R''_{ab}(c,c)=-\frac{\alpha(N)}{c},\ S''_{ab}(c,c)=-\frac{\alpha(S)}{c},$$

then

$$p(c)=k\cdot c^{[1+8\cdot\alpha(N)]/n-8\cdot[\alpha(R)-\alpha(S)]/n^2-m}+h,$$

where k and h are arbitrary constants.

Remark 109. We can deduce all the known special cases published on this subject.

Remark 110. In Kim (1999), it was also proved that in some cases the conditions are sufficient, thus the author obtained the integral representation of the pre-means given by (2.18) and (2.19).

4.8 ESTIMATIONS OF SOME INTEGRAL MEANS

Using (4.1)–(4.3), the arithmetic–geometric mean $\mathcal{A} \otimes \mathcal{G} = \mathcal{M}$ can be represented as

$$\mathcal{M}(a,b) = M_{e_{-1/2},e_1,e_2}(a,b) = e_{1/2} \circ e_{-2}(f(a,b;e_{-1/2},e_1,e_2))$$
$$= \frac{a\pi}{2K\left(\sqrt{1-\left(\frac{b}{a}\right)^2}\right)}.$$

Similarly, in the Chapter 1 Section 1.8 we saw that the perimeter of an ellipse is given by

$$L(a,b) = \int_0^{2\pi} \sqrt{a^2\cos^2\theta + b^2\sin^2\theta}d\theta$$
$$= 2\pi f(a,b;e_{1/2},e_1,e_2) = 2\pi M_{e_{1/2},e_1,e_2}(a,b)$$
$$= 4a \cdot E\left(\sqrt{1-\left(\frac{b}{a}\right)^2}\right).$$

The mean

$$L/2\pi = M_{e_{1/2},e_1,e_2} = \mathcal{U}$$

was named in Chu et al. (2011) as **Toader mean** with reference to the paper Toader (1998), and it was studied also in Chu et al. (2012).

The above two integral means, being related to the elliptic integrals, are the most compared and estimated by other means.

The AGM was evaluated from below in Carlson and Vuorinen (1991) and from above in Borwein and Borwein 1(993) by logarithmic means, giving

$$\mathcal{L} < \mathcal{M} < \mathcal{L}_{3/2}.$$

Then in Vamanamurthy and Vuorinen (1994), the optimal estimation

$$\mathcal{P}_0 < \mathcal{M} < \mathcal{P}_{1/2}$$

is proved, while in Chu and Wang (2012a), another optimal estimation

$$\mathcal{S}_{1/2} < \mathcal{M} < \mathcal{S}_1$$

is given.

Among the approximations of the perimeter of an ellipse analyzed in Almkvist and Berndt (1988) going from that of Kepler from 1609 up to that of Jacobsen from 1985, let us mention that of Muir from 1883. He used for $L(a, b)$ the approximation by $2\pi \cdot \mathcal{P}_{3/2}(a, b)$.

In Vuorinen (1998), it is conjectured that this is an approximation from below, which was proved to be true in Barnard et al. (2000). In Anderson et al. (2007), it is stated that in the previous paper the upper estimation $2\pi \mathcal{P}_2(a, b)$ is also given, but the reference seems to be wrong. Anyway, in Alzer and Qiu (2004), the optimal estimation

$$\frac{\pi}{2}\mathcal{P}_{3/2}(1, r') < \mathcal{E}(r) < \frac{\pi}{2}\mathcal{P}_{\frac{\ln 2}{\ln(\pi/2)}}(1, r')$$

was proved, where $r' = \sqrt{1 - r^2}$ gives the optimal estimation

$$\mathcal{P}_{3/2} < \mathcal{U} < \mathcal{P}_{\ln 2/\ln(\pi/2)}.$$

Other optimal estimations of the mean $\mathcal{U}$ were obtained in Chu et al. (2011):

$$\mathcal{T}_{\frac{\sqrt{3}}{4}} < \mathcal{U} < \mathcal{T}_{\frac{1}{2}},$$

in Chu and Wang (2012):

$$\mathcal{S}_1 < \mathcal{U} < \mathcal{S}_{3/2},$$

respectively in Chu and Wang (2012a):

$$\mathcal{C}_1 < \mathcal{U} < \mathcal{C}_{5/4}.$$

Other types of estimations were obtained in Song et al. (2013).

Bibliography

Abramovich, S., Pečarić, J., 2000. Functional equalities and some mean values. J. Math. Anal. Appl. 250, 181–186.

Aczél, J., 1948. On mean values. Bull. Amer. Math. Soc. 54, 392–400.

Aczél, J., Losonczi, L., Páles, Z., 1987. The behavior of comprehensive classes of means under equal increments of their variables. In: Walter, W. (Ed.), Gen. Ineq. 5: 5th International Conference on General Inequalities. Oberwolfach, Germany, May 4–10, 1986. In: Internat. Series Numer. Math., vol. 80, pp. 459–461.

Allasia, G., 1969–1970. Su una classe di algoritmi iterativi bidimensionali. Rend. Sem. Mat. Univ. Politech. Torino 29, 269–296.

Allasia, G., 1970–1971. Relazioni tra una classe di algoritmi iterative bidimensionali ed una di equazioni differenziali. Rend. Sem. Mat. Univ. Politech. Torino 30, 187–207.

Allasia, G., 1971–1972. Alcune generalizzazioni dell'algoritmo della media aritmetico-armonica. Rend. Sem. Mat. Univ. Politech. Torino 31, 197–221.

Allasia, G., 1972. Sul calcolo numerico di due prodotti infiniti. Inst. Calc. Num. Univ. Politech. Torino, Quaderno 2, 1–19.

Allasia, G., 1983. Algoritmi iterative del tipo delle medie. Proprietà generali. Accad. Pelor. Pericol. Messina 61, 87–111.

Allasia, G., Bonardo, F., 1980. On the numerical evaluation of two infinite products. Math of Comp. 35, 917–931.

Almkvist, G., 1978. Aritmetisk-geometrisko medelvärdet och ellipsens båglängd. Nordisk Mat. Tidskr. 25–26, 121–130.

Almkvist, G., Berndt, B., 1988. Gauss, Landen, Ramanujan, the arithmetic–geometric mean, ellipses, and the Ladies Diary. Amer. Math. Monthly 95 (7), 585–608.

Alzer, H., 1993. Best possible estimates for special mean values. Zb. Rod. Prirod. Mat. Fak. Ser. Mat. 23 (1), 331–346 (in German).

Alzer, H., Qiu, S.-L., 2004. Monotonicity theorems and inequalities for the complete elliptic integrals. J. Comput. Appl. Math. 172, 289–312.

Anderson, G., Vamanamurthy, M.K., Vuorinen, M., 2007. Topics in special functions. arXiv:0712.3856v1 [math.CA].

Andreoli, G., 1957. Aspetto gruppale e funzionale delle medie. Giornale Matem. Battaglini 85 (5), 12–30.

Andreoli, G., 1957a. Medie e loro processi iterative. Giornale Matem. Battaglini 85 (5), 52–79.

Antoine, C., 1998. Les Moyennes. Presses Unversitaires de France, Paris.

Arazy, J., Claesson, T., Janson, S., Peetre, J., 1985. Means and their iterations. In: Proc. Nineteenth Nordic Congr. Math. Reykjavik, 1984. In: Visindafél Isl., vol. XLIV. Icel. Math. Soc., Reykjavik, pp. 191–212.

Aumann, G., 1935. Über den Mischalgorithmus bei analytischen Mittelwerten. Math. Zeit. 39, 625–629.

Baják, S., Páles, Z., 2009. Invariance equation for generalized quasi-arithmetic means. Aequationes Math. 77, 133–145.

Baják, S., Páles, Z., 2009a. Computer aided solution of the invariance equation for two-variable Gini means. Comput. Math. Appl. 58, 334–340.

Baják, S., Páles, Z., 2012. Invariance Equation for Two-Variable Stolarsky Means. PhD thesis. University of Debrecen, Faculty of Science and Technology, Hungary.

Bajraktarević, M., 1958. Sur une équation fonctionelle aux valeurs moyennes (On a functional equation with mean values). Glas. Mat. Ser. III 13, 243–248 (in French).

Barna, B., 1934. Ein Limessatz aus der Theorie des arithmetisch-geometrischen Mittels (A set of theorems of the arithmetic–geometric mean). J. Reine Angew. Math. 172, 86–88 (in German).

Barna, B., 1939. Zur elementaren Theorie des arithmetisch-geometrischen Mittels (On the elementary theory of the arithmetic–geometric mean). J. Reine Angew. Math. 178, 129–134 (in German).

Barnard, R.W., Pearce, K., Richards, K., 2000. An inequality involving the generalized hypergeometric function and the arc length of an ellipse. SIAM J. Math. Anal. 31 (3), 693–699.

Bauer, H., 1986. A class of means and related inequalities. Manuscripta Math. 55, 199–212.

Baxter, L., 1981. Are π, e and $\sqrt{2}$ equally difficult to compute? Amer. Math. Monthly 88 (1), 50–51.

Beke, E., 1927. Eine Mittelwert (A mean value). Z. Math. Unterricht. 58, 325–326 (in German).

Bellman, R., 1956–1957. On the arithmetic–geometric mean inequality. Math. Student 25, 233–234.

Berggren, L., Borwein, J.M., Borwein, P.B. (Eds.), 1997. Pi: A Source Book. Springer-Verlag, New York.

Bernoulli, J., 1744. Opera I. Geneva.

Berrone, L.R., Moro, J., 1998. Lagrangian means. Aequationes Math. 55 (3), 217–226.

Besenyei, A., 2012. On the invariance equation for Heinz means. Math. Ineq. Appl. 12 (4), 973–979.

Bhatia, R., 2006. Interpolating the arithmetic–geometric mean inequality and its operator version. Linear Algebra Appl. 413, 355–363.

Bierens de Haan, D., 1867. Nouvelles tables d'intégrales définies. Leide: P. Engels, Amsterdam.

Błasińska-Lesk, J., Głazowska, D., Matkowski, J., 2003. An invariance of the geometric mean with respect to Stolarsky mean-type mappings. Result. Math. 43, 42–55.

Borchardt, C.W., 1861. Über das arithmetisch-geometrische Mittel. J. Reine Angew. Math. 58, 127–134.

Borchardt, C.W., 1876. Über das arithmetisch-geometrische Mittel aus vier Elementen. Monatsh. Akad. Wiss. Berlin, 611–621.

Borchardt, C.W., 1881. Sur deux algorithms analogues à celui de la moyenne aritmético-géometrique de deux elements. In: Cremona, L. (Ed.), In memoriam Dominicci Chelini, Collectaanea Mathemaatica exc. U. Hoepli, Milano.

Borwein, D., Borwein, P.B., 1983. Problem 83-12. SIAM Rev. 25 (3), 401.

Borwein, J.M., 1996. Solution of the Problem 10281. A cubic relative of the AGM. Amer. Math. Monthly 103 (2), 181–183.

Borwein, J.M., Borwein, P.B., 1983a. A very rapidly convergent product expansion for π. BIT Numerical Mathematics 23 (4), 538–540.

Borwein, J.M., Borwein, P.B., 1984. The arithmetic–geometric mean and fast computation of elementary functions. SIAM Rev. 26 (3), 351–366.

Borwein, J.M., Borwein, P.B., 1987. Pi and the AGM – a Study in Analytic Number Theory and Computational Complexity. John Wiley & Sons, New York.

Borwein, J.M., Borwein, P.B., 1987a. The way of all means. Amer. Math. Monthly 94 (6).

Borwein, J.M., Borwein, P.B., 1989. On the mean iteration $(a, b) \longleftarrow ((a + 3b)/4, (\sqrt{ab} + b)/2)$. Math. Comp. 53, 311–326.

Borwein, J.M., Borwein, P.B., 1991. A cubic counterpart of Jacobi's identity and the AGM. Trans. Amer. Math. Soc. 323, 691–701.

Borwein, J.M., Borwein, P.B., 1993. Inequalities for compound mean iterations with logarithmic asymptotes. J. Math. Anal. Appl. 177, 572–582.

Borwein, P.B., 1991. Quadratically converging rational mean iterations. J. Math. Anal. Appl. 154, 361–376.

Braden, B., Danloy, B., Schmidt, F., 1992. Solution of the problem 91-17. SIAM Rev. 34 (4), 653–654.

Brenner, J.L., Mays, M.E., 1987. Some reproducing identities for families of mean values. Aequationes Math. 33, 106–113.

Bullen, P.S., 1998. A Dictionary of Inequalities. Addison–Wesley Longman, London.

Bullen, P.S., 2003. Handbook of Means and Their Inequalities. Kluwer Academic Publishers, Dordrecht, Boston, London.

Bullen, P.S., Mitrinović, D.S., Vasić, P.M., 1988. Means and Their Inequalities. D. Reidel Publ. Comp., Dordrecht.

Bullett, S., 1991. Dynamics of arithmetic–geometric mean. Topology 30, 171–190.

von Bültzingslöven, W., 1933. Iterative algebraische Algorithmen. Mitt. Math. Sem. Giessen 23, 1–72.

Burai, P., 2007. A Matkowski–Sutô type equation. Publ. Math. Debrecen 70 (1–2), 233–247.

Buzano, M.L., 1965–1966. Sulle curve uniti di talune transformazioni puntuali. Rend. Sem. Mat. Univ. Politech. Torino 25, 193–210.

Carlson, B.C., 1971. Algorithms involving arithmetic and geometric means. Amer. Math. Monthly 78 (5), 496–505.

Carlson, B.C., 1972. The logarithmic mean. Amer. Math. Monthly 79, 615–618.

Carlson, B.C., 1975. Invariance of an integral average of a logarithm. Amer. Math. Monthly 82, 379–382.

Carlson, B.C., Gustafson, J.T., 1983. Total positivity of mean values and hypergeometric functions. SIAM J. Math. Anal. 14 (2), 389–395.

Carlson, B.C., Vuorinen, M., 1991. SIAM Rev. 33 (4), 655.

Chajoth, Z., 1932. Heronische Näherungsbrüche. Jber. Deutsch. Math.-Verein. 42, 130–135.

Chisini, O., 1929. Sul concetto di media. Period. Mat. 4, 106–116.

Chu, Y.-M., Long, B.-Y., 2013. Bounds of the Neuman–Sándor mean using power and identric mean. Abstract Appl. Anal. 2013, 832591. http://dx.doi.org/10.1155/2013/832591.

Chu, Y.-M., Wang, M.-K., Qiu, Y.-F., 2011. Sharp generalized Seiffert mean bounds for Toader mean. Abstract Appl. Anal. 2011, 605259. http://dx.doi.org/10.1155/2011/605259.

Chu, Y.-M., Wang, M.-K., 2012. Optimal Lehmer mean bounds for the Toader mean. Result. Math. 61 (3–4), 223–229.

Chu, Y.-M., Wang, M.-K., 2012a. Inequalities between arithmetic–geometric, Gini, and Toader mean. Abstr. Appl. Anal. 2012, 830585. http://dx.doi.org/10.1155/2012/830585.

Chu, Y.-M., Wang, M.-K., Qiu, S.-L., 2012. Optimal combinations bounds of root-square and arithmetic means for Toader mean. Proc. Indian Acad. Sci. Math. Sci. 122 (1), 41–51.

Ciorănescu, N., 1936. On the arithmetic–geometric mean. Bul. Şed. Soc. Rom. St. Bucureşti 1, 17 (in Romanian).

Ciorănescu, N., 1936a. L'itération des fonctions de moyen. Bull. Math. Soc. Sci. Math. R. S. Roumanie 38, 71–74.

Claesson, T., Peetre, J., 1990. On an algorithm considered by Stieltjes. J. Math. Anal. Appl. 150 (2), 481–493.

Costin, I., 2003. Series expansion of means. In: Proceedings of the International Symposium Specialization, Integration and Development, Section: Quantitative Economics. Babeş-Bolyai University Cluj-Napoca, Romania, pp. 115–122.

Costin, I., 2004. Generalized inverses of means. Carpathian J. Math. 20 (2), 169–175.

Costin, I., 2004a. Acceleration of the convergence of Archimedean double sequences. In: Gavrea, I., Ivan, M. (Eds.), Mathematical Analysis and Approximation Theory RoGer 2004. Cluj-Napoca, Romania. Mediamira Science Publisher, pp. 81–88.

Costin, I., 2004b. Complementariness with respect to power means. Automat. Comput. Appl. Math. 13 (1), 69–77.

Costin, I., 2007. Invariance in the class of weighted power means. In: 9th International Symposium on Symbolic and Numerical Algorithms for Scientific Computing SYNASC 2007. Timişoara, Romania. IEEE Computer Society Conference Publishing Services, Los Alamos, California, pp. 131–133.

Costin, I., 2007a. Power means. Automat. Comput. Appl. Math. 16 (2), 25–31.

Costin, I., 2008. Complementaries of Greek means with respect to the logarithmic mean. In: 10th International Symposium on Symbolic and Numerical Algorithms for Scientific Computing SYNASC 2008. Timişoara, Romania. IEEE Computer Society Conference Publishing Services, Los Alamos, California, pp. 81–85.

Costin, I., 2010. Invariance of the logarithmic mean with respect to the family of Stolarsky means. Automat. Comput. Appl. Math. 19 (2), 37–44.

Costin, I., Toader, G., 2003. A weighted Gini mean. In: Proceedings of the International Symposium Specialization, Integration and Development, Section: Quantitative Economics. Babeş-Bolyai University Cluj-Napoca, Romania, pp. 109–114.

Costin, I., Toader, G., 2004. On the rate of convergence of Archimedean double sequences. Creative Math. 13, 1–4.

Costin, I., Toader, G., 2004a. Archimedean double sequences. In: Gavrea, I., Ivan, M. (Eds.), Mathematical Analysis and Approximation Theory RoGer 2004. Mediamira Science Publisher, Cluj-Napoca, Romania, pp. 89–98.

Costin, I., Toader, G., 2005. Generalized inverses of Lehmer means. Automat. Comput. Appl. Math. 14 (1), 111–117.

Costin, I., Toader, G., 2005a. Generalized inverses of power means. General Math. 13, 61–70.

Costin, I., Toader, G., 2006. Gaussian double sequences. Automat. Comput. Appl. Math. 15 (1), 103–109.

Costin, I., Toader, G., 2006a. Inequalities for compound means. In: Cho, Y.J., Kim, J.K., Dragomir, S.S. (Eds.), Inequal. Th. Appl., vol. 5. Nova Sci. Publ., Inc., New York.

Costin, I., Toader, G., 2006b. Generalized inverses of Gini means. Automat. Comput. Appl. Math. 15 (1), 111–115.

Costin, I., Toader, G., 2008. Invariance in the class of weighted Lehmer means. J. Ineq. Pure Appl. Math. 9 (2). Art. 54, 7 pp. https://www.emis.de/journals/JIPAM/article986.html?sid=986.

Costin, I., Toader, G., 2008a. Complementariness with respect to the identric mean. Automat. Comput. Appl. Math. 17 (3), 409–415.

Costin, I., Toader, G., 2012. A nice separation of some Seiffert-type means by power means. Internat. J. Math. Math. Sci. 2012, 430692. http://dx.doi.org/10.1155/2012/430692.

Costin, I., Toader, G., 2012a. A separation of some Seiffert-type means by power means. Rev. Anal. Num. Th. Approx. 41 (2), 125–129.

Costin, I., Toader, G., 2012b. Invariance of a weighted power mean in the class of weighted Gini means. Automat. Comput. Appl. Math. 21 (1), 35–43.

Costin, I., Toader, G., 2013. Optimal evaluations of some Seiffert-type means by power means. Appl. Math. Comput. 219, 4745–4754.

Costin, I., Toader, G., 2013a. Invariance of a weighted Lehmer mean in the class of weighted Gini means. Automat. Comput. Appl. Math. 22 (1), 89–101.

Costin, I., Toader, G., 2013b. Some optimal evaluations of the logarithmic mean. Automat. Comput. Appl. Math. 22 (1), 103–112.

Costin, I., Toader, G., 2014. Optimal estimations of some Seiffert-type means by Lehmer means. RGMIA Research Report Collection 17. Art. 24, 8 pp.

Costin, I., Toader, G., 2014a. Optimal estimations of Seiffert-type means by some special Gini means. In: Gerdt, V.P., Koepf, W., Seiler, W.M., Vorozhtsov, E.V. (Eds.), Computer Algebra in Scientific Computing – 16th International Workshop, CASC 2014. Warsaw, Poland, September 8–12, 2014. In: Lecture Notes Comp. Sci., vol. 8660. Springer International Publishing, pp. 86–99.

Costin, I., Toader, G., 2014b. Invariance in the family of weighted Gini means. In: Rassias, T.M. (Ed.), Handbook of Functional Equations: Functional Inequalities. In: Springer Optim. Appl., vol. 95, pp. 105–127.

Costin, I., Toader, S., 2009. Complementariness with respect to the logarithmic mean. Math. Ineq. Appl. 12 (4), 791–798.

Cox, D.A., 1984. The arithmetic–geometric mean of Gauss. L'Enseign. Math. 30, 275–330.

Cox, D.A., 1985. Gauss and the arithmetic–geometric mean. Notices of the Amer. Math. Soc. 32 (2), 147–151.

Czinder, P., Páles, Z., 2000. A general Minkowski-type inequality for two varianle Gini means. Publ. Math. Debrecen 57 (1–2), 203–216.

Czinder, P., Páles, Z., 2003. Minkowski-type inequalities for two variable Stolarsky means. Acta Sci. Math. (Szeged) 69, 27–47.

Czinder, P., Páles, Z., 2006. Some comparison inequalities for Gini and Stolarsky means. Math. Inequal. Appl. 9 (4), 607–616.

Daróczy, Z., 2005. Functional equations involving means and Gauss compositions of means. Nonlinear Anal. 63, 417–425.

Daróczy, Z., Hajdu, G., Ng, C.T., 2003. An extension theorem for a Matkowski–Sutô problem. Colloq. Math. 95 (2), 153–161.

Daróczy, Z., Maksa, G., Páles, Z., 2005. Functional equations involving means and their Gauss compositions. Proc. Amer. Math. Soc. 134 (2), 521–530.

Daróczy, Z., Páles, Z., 2001. On means that are both quasi-arithmetic and conjugate arithmetic. Acta Math. Hungar. 90, 271–282.

Daróczy, Z., Páles, Z., 2002. Gauss-composition of means and the solution of the Matkowski–Sutô problem. Publ. Math. Debrecen 61 (1–2), 157–218.

Daróczy, Z., Páles, Z., 2003. On functional equations involving means. Publ. Math. Debrecen 62 (3–4), 363–377.

Daróczy, Z., Páles, Z., 2003a. The Matkowski–Sutô problem for weighted quasi-arithmetic means. Acta Math. Hung. 100 (3), 237–243.

von Dávid, L., 1907. Theorie der Gauss'schen verallgemeinerten und speziellen arithmetisch-geometrisches Mittels. Math.-Naturw. Berichte aus Ungarn 25, 153–177.

von Dávid, L., 1909. Sur une application des fonctions modulaires à la théorie de la moyenne aritmético-geometric. Math.-Naturw. Berichte aus Ungarn 27, 164–171.

von Dávid, L., 1913. Zur Gauss'schen Theorie der Modulfunktion. Rend. Circ. Mat. Palermo 25, 82–89.

von Dávid, L., 1928. Arithmetisch-geometrisches Mittel und Modulfunktion. J. Reine Angew. Math. 159, 154–170.

Dmitrieva, O.M., Malozemov, V.N., 1997. AGH means. Vestnik St. Petersburg Univ. Math. 30 (2), 56–57.

Domsta, J., Matkowski, J., 2006. Invariance of the arithmetic mean with respect to special mean-type mappings. Aequationes Math. 71, 70–85.

Euler, L., 1774. Nova series infinita maxime convergens perimetrum ellipsis exprimens. Novi Comm. Acad. Sci. Petropolitanae 18, 71–84. In: Opera Omnia. In: Series Prima, vol. XX. B. G. Teubner, Leipzig, Berlin, 1912, pp. 357–370.

Euler, L., 1786. De miris proprietatibus curvae elasticae sub aequatione $y = \int x^2/\sqrt{1-x^4}dx$ contentae. Novi Comm. Acad. Sci. Petropolitinae II, 34–61. In: Opera Omnia. In: Series Prima, vol. XXI. B. G. Teubner, Leipzig, Berlin, 1913, pp. 91–118.

Everett, C., Metropolis, N., 1971. A generalization of the Gauss limit for iterated means. Adv. in Math. 7, 197–300.

Eymard, P., Lafon, J.-P., 2004. The Number π. Amer. Math. Soc., Providence, Rhode Island.

Faragó, T., 1951. Über das aritmetisch-geometrisch Mittel. Publ. Math. Debrecen 2, 150–156.

Farhi, B., 2010. Algebraic and topological structures on the set of means functions and generalization of the AGM mean. arXiv:1002.4757 [math.NT]. 23 pp.

Foster, D.M.E., Phillips, G.M., 1983. The approximation of certain functions by compound means. In: Singh, S.P., Burry, J.W.H., Watson, B. (Eds.), Approximation Theory and Spline Functions. In: NATO ASI Series, vol. 136, pp. 89–95.

Foster, D.M.E., Phillips, G.M., 1984. A generalization of the Archimedean double sequence. J. Math. Anal. Appl. 110 (2), 575–581.

Foster, D.M.E., Phillips, G.M., 1984a. The arithmetic-harmonic mean. Math. of Comp. 42 (165), 183–191.

Foster, D.M.E., Phillips, G.M., 1985. General compound means. In: Approximation Theory and Applications. St. John's, Nfld., 1984. In: Res. Notes in Math., vol. 133. Pitmann, Boston, Mass., London, pp. 56–65.

Foster, D.M.E., Phillips, G.M., 1986. Double mean processes. Bull. Inst. Math. Appl. 22 (11–12), 170–173.

Fricke, R., 1901–1921. Geometrisch'Entwicklungen über das aritmetisch-geometrisch Mittel. In: Encykl. Math. Wiss., II, vol. 2, pp. 222–225.

Frisby, E., 1879. On the aritmetico-geometrical mean. In: The Analyst. J. Pure Appl. Math. 6, 10–14.

Fuchs, W., 1972. Das arithmetisch-geometrische Mittel in den Untersuchungen von Carl Friedrich Gauss. Gauss-Gesellschaft Göttingen, Mittelungen 9, 14–38.

Ganelius, T.H., Hayman, W.K., Newman, D.J., 1982. Lectures on Approximation and Value Distribution. University of Montreal Press.

Gatteschi, L., 1966. Su una generalizzazione dell'algoritmo iterativo del Borchardt. Mem. Acad. Sci. Torino 4 (4), 1–18.

Gatteschi, L., 1969–1970. Procedimenti iterative per il calcolo numerico di due prodotti infiniti. Rend. Sem. Mat. Torino 29, 187–201.

Gatteschi, L., 1982. Il contributo di Guido Fubini agli algoritmi iterativi. Atti Accad. Sci. Torino Cl. Sci. Fis. Mat. Natur., Suppl. 115, 61–70.

Gauss, C.F., 1800. Nachlass. Arithmetisch-geometrisches Mittel (1800). In: Werke, vol. 3. Königlichen Gesell. Wiss, Göttingen, pp. 361–403.

Gauss, C.F., 1876–1927. Werke. Königlichen Gesell. Wiss., Göttingen, Leipzig.

Georgakis, C., 2002. On the inequality for the arithmetic and geometric means. Math. Ineq. Appl. 5, 215–218.

Geppert, H., 1928. Zur Theorie des arithmetisch-geometrischen Mittels. Math. Ann. 99, 162–180.

Geppert, H., 1932. Über iterative Algorithmen I. Math. Ann. 107, 387–399.

Geppert, H., 1933. Über iterative Algorithmen II. Math. Ann. 108, 197–207.

Gini, C., 1958. Le Medie. Unione Tipografico Torinese, Milano.

Giroux, A., Rahman, Q.I., 1982. Research problems in function theory. Ann. Sc. Math. Québec 6 (1), 71–79.

Głazowska, D., Jarczyk, W., Matkowski, J., 2002. Arithmetic mean as a linear combination of two quasi-arithmetic means. Publ. Math. Debrecen 61 (3–4), 455–467.

Głazowska, D., Matkowski, J., 2007. An invariance of geometric mean with respect to Lagrangian means. J. Math. Anal. Appl. 331, 1187–1199.

Gosiewski, W., 1909. O sredniej arytmetycznej o prawie Gauss'a prawdopobieństwa. Sprawozdanie Towarzystwa Nauk. Warszaw. 2, 11–17.

Gould, H.W., 1974. Coefficient identities for powers of Taylor and Dirichlet series. Amer. Math. Monthly 81, 3–14.

Gould, H.W., Mays, M.E., 1984. Series expansions of means. J. Math. Anal. Appl. 101 (2), 611–621.

Gradshteyn, I.S., Ryzhik, I.M., 1994. Tables of Integrals, Series and Products. Acad. Press, New York.

Grayson, D.R., 1989. The arithogeometric mean. Arch. Math. (Basel) 52, 507–512.

Gustin, W.S., 1947. Gaussian means. Amer. Math. Monthly 54, 332–335.

Hardy, G.H., Littlewood, J.E., Pólya, G., 1934. Inequalities. Cambridge.

Haruki, H., 1991. New characterizations of the arithmetic–geometric mean of Gauss and other well-known mean values. Publ. Math. Debrecen 38 (3–4), 323–332.

Haruki, H., Rassias, T.M., 1995. A new analogue of Gauss' functional equation. Internat. J. Math. Math. Sci. 18 (4), 749–756.

Haruki, H., Rassias, T.M., 1996. New characterizations of some mean-values. J. Math. Anal. Appl. 202 (1), 333–348.

Hästö, P.A., 2004. Optimal inequalities between Seiffert's means and power means. Math. Ineq. Appl. 7 (1).

Heath, T., 1949. A History of Greek Mathematics. Oxford University Press, Oxford.

Heck, A., 2003. Introduction to Maple, 3rd ed. Springer-Verlag, New York, Berlin.

Heinrich, H., 1981. Eine Verallgemeinerung des arithmetisch-geometrischen Mittels. Z. Angew. Math. Mech. 61, 265–267.

Hettner, G., 1880. Zur Theorie der arithmetisch-geometrisch Mittel. J. Reine Angew. Math. 89, 221–246.

Hofsommer, D.J., van de Riet, R.P., 1962. On the Numerical Calculation of Elliptic Integrals of the 1st and 2nd Kind and the Elliptic Functions of Jacobi. Report TW94. Stichtung Math. Centrum, Amsterdam.

Ivory, J., 1796. A new series for the rectification of the ellipses, togethere with some observations on the evolution of the formula $(a^2 + b^2 - 2ab\cos\varphi)^n$. Trans. Royal Soc. Edinburgh 4, 177–190.

Jacobi, C.C.J., 1881. Fundamenta nova theoriae functionum ellipticorum. In: Gesammelte Werke. G. Reimer, Berlin, pp. 49–239.

Jagers, A.A., 1994. Solution of problem 887. Nieuw Arch. Wisk. 12, 230–231.

Janous, W., 2001. A note on generalized Heronian means. Math. Ineq. Appl. 4 (3), 369–375.

Jarczyk, J., 2007. Invariance in the class of weighted quasi-arithmetic means with continuous generators. Publ. Math. Debrecen 71, 279–294.

Jarczyk, J., 2009. Invariance of quasi-arithmetic means with function weights. J. Math. Anal. Appl. 353, 134–140.

Jarczyk, J., 2010. Invariance in a class of Bajraktarevic' means. Nonlinear Anal., Th. Methods Appl., Ser. A, Theory Methods 72 (5A), 2608–2619.

Jarczyk, J., Matkowski, J., 2006. Invariance in the class of weighted quasi-arithmetic means. Ann. Polon. Math. 88 (1), 39–51.

Jecklin, H., 1948. Der Begriff des mathematischen Mittelwertes und die Mittelwertformeln. Viertdjschr. Naturforsch. Ges. Zürich 93, 35–41.

Jecklin, H., 1949. Versuch einer Systematik des mathematischen Mittelwertbegriffs. Comment. Math. Helv. 22, 260–270.

Jecklin, H., 1953. Trigonometrische Mittelwerte. El. Math. 8, 3.

Jia, G., Cao, J., 2003. A new upper bound of the logarithmic mean. J. Ineq. Pure Appl. Math. 4 (4). Art. 80.

Kahlig, P., Matkowski, J., 1997. On the composition of homogeneous quasi-arithmetic means. J. Math. Anal. Appl. 216, 69–85.

Kämmerer, F., 1925. Ein arithmetisch-geometrisches Mittel. Jber. Deutsch. Math.-Verein. 34, 87–88.

Karamata, J., 1960. Sur quelques problèmes posés par Ramanujan. J. Indian Math. Soc. (N. S.) 24, 343–365.

Kellog, O.D., 1929. Foundation of Potential Theory. Springer Verlag.

Kepler, J., 1860. Opera Omnia, vol. 3. Astronomia Nova, Heyder & Zimmer, Frankfurt.

Kim, Y.-H., 1999. On some further extensions of the characterizations of mean values by H. Haruki and T.M. Rassias. J. Math. Anal. Appl. 235 (2), 598–607.

Kim, Y.-H., Rassias, T.M., 2002. Properties of some mean values and functional equations. Panamer. Math. J. 12 (1), 65–74.

Kim, Y.-H., Ume, J.S., 2000. New characterizations of well-known mean-values II. Far East J. Math. Sc. 2 (3), 453–461.

Kuznecov, V.M., 1969. On the work of Gauss and Hungarian mathematicians on the theory of the aritmetico-geometric mean. Rostov-na-Don Gos. Univ.-Nauch. Soobsh., 110–115.

Lagrange, J.L., 1784–1785. Sur une nouvelle méthode de calcul integral pour les différentielles affectées d'un radical carré sous lequel la variable ne passe pas le quatriéme degrè. In: Mem. L'Acad. Roy. Sci. Turin, vol. 2. In: Oeuvres, vol. 2. Gauthier-Villars, Paris, 1868, pp. 251–312.

Landen, J., 1771. A disquisition concerning certain fluents, which are assignable by the arcs of the conic sections, wherein are investigated some new and useful theorems for computing such fluents. Philos. Trans. Royal Soc. London 61, 298–309.

Leach, E.B., Sholander, M.C., 1983. Extended mean values II. J. Math. Anal. Appl. 92, 207–223.

Legendre, A.M., 1825. Traité des fonctions elliptiques. Huzard-Courcier, Paris.

Lehmer, D.H., 1971. On the compounding of certain means. J. Math. Anal. Appl. 36, 183–200.

Lin, T.P., 1974. The power mean and the logarithmic mean. Amer. Math. Monthly 81, 879–883.

Liu, Z., 2001. On new generalizations of certain mean values by H. Haruki and T.M. Rassias. Math. Sci. Hot-Line 5 (5), 15–23.

Liu, Z., 2001a. On Gauss type functional equations and mean values by H. Haruki and T.M. Rassias. Studia Univ. Babeş-Bolyai, Math. 46 (4), 75–87.

Liu, Z., 2004. Compounding of Stolarsky means. Soochow J. Math. 30 (2), 149–163.

Lohnsein, T., 1888. Zur Theorie des arithmetisch-geometrisch Mittels. Z. Math. Phys. 33, 129–136.

Lohnsein, T., 1888a. Über das harmonische-geometrische Mittel. Z. Math. Phys. 33, 316–318.

MacLaurin, C., 1742. A Treatise of Fluxions in Two Books. T.W. and T. Ruddimans, Edinburgh.

Makó, Z., Páles, Z., 2009. The invariance of the arithmetic mean with respect to generalized quasi-arithmetic means. J. Math. Anal. Appl. 353, 8–23.

Mathieu, J.J., 1879. Note relative à l'approximation des moyens géométriques par des séries de moyens arithmétiques et de moyens harmoniques. Nouv. Ann. Math. XVIII, 529–531.

Matkowski, J., 1999. Invariant and complementary quasi-arithmetic means. Aequationes Math. 57, 87–107.

Matkowski, J., 1999a. Iteration of mean-type mappings and invariant means. Ann. Math. Sil. 13, 211–226.

Matkowski, J., 2002. On invariant generalized Beckenbach–Gini means. In: Daróczy, Z., Páles, Z. (Eds.), Functional Equations – Results and Advances. In: Adv. Math., vol. 3. Kluwer Acad. Publ., Dordrecht, pp. 219–230.

Matkowski, J., 2002a. On iteration semigroups of mean-type mappings and invariant means. Aequationes Math. 64, 297–303.

Matkowski, J., 2004. Remarks on some problems of Th.M. Rassias. Aequationes Math. 67, 197–200.

Matkowski, J., 2005. Lagrangian mean-type mappings for which the arithmetic mean is invariant. J. Math. Anal. Appl. 309, 15–24.

Matkowski, J., 2006. On iteration of means and functional equations. In: Förg-Rob, W. (Ed.), Iteration Theory (ECIT '04). In: Grazer Math. Ber. Bericht, vol. 350, pp. 184–201.

Matkowski, J., 2009. Iterations of the mean-type mappings. In: Sharkovsky, A.N., Sushko, I.M. (Eds.), Iteration Theory (ECIT '08). In: Grazer Math. Ber. Bericht, vol. 354, pp. 158–179.

Matkowski, J., 2011. Quotient mean, its invariance with respect to a quasi-arithmetic mean-type mapping, and some applications. Aequat. Math. 82, 247–253.

Mays, M.E., 1983. Functions which parametrize means. Amer. Math. Monthly 90, 677–683.

Miel, G., 1983. Of calculation past and present: the Archimedean algorithm. Amer. Math. Monthly 90 (1), 17–35.
Mohr, E., 1935. Über die Funktionalgleichung des arithmetisch-geometrisch. Mittel. Math. Nachr. 10, 129–133.
Moskovitz, D., 1933. An alignment chart for various means. Amer. Math. Monthly 40, 592–596.
Muntean, I., Vornicescu, N., 1993–1994. The arithmetic–geometric mean and the computation of elliptic integrals. Lucrarile Semin. Didactica Matematica, (Cluj-Napoca) 10, 57–76 (in Romanian).
Myrberg, P.J., 1958. Sur une generalization de la moyenne arithmétique- géométrique de Gauss. C. R. Acad. Sc. Paris 246 (23), 3201–3204.
Myrberg, P.J., 1958a. Eine Verallgemeinerung des arithmetisch-geometrischen Mittels. Annales Acad. Sc. Fennicae I, Math. 253, 3–19.
Neuman, E., 1994. The weighted logarithmic mean. J. Math. Anal. Appl. 188 (3), 885–900.
Neuman, E., 2013. A one-parameter family of bivariate means. J. Math. Inequal. 7, 399–412.
Neuman, E., 2014. On some means derived from the Schwab–Borchardt mean. J. Math. Inequal. 8, 171–183.
Neuman, E., Páles, Z., 2003. On comparison of Stolarsky and Gini means. J. Math. Anal. Appl. 278, 274–284.
Neuman, E., Sándor, J., 2003. On the Schwab–Borchardt mean. Math. Panonica 14 (2), 253–266.
Newman, D.J., 1982. Rational approximation versus fast computer methods. In: Lectures on Approximation and Value Distribution. Presses de l'Univ. Montréal, pp. 149–174.
Newman, D.J., 1985. A simplified version of the fast algorithms of Brent and Salamin. Math. Comp. 44, 207–210.
Nikolaev, A.N., 1925. Extraction of square roots and cubic roots with the help of calculators. Tashkent. Bull. Univ. 11, 65–74.
Nowicki, T., 1998. On the arithmetic and harmonic means. In: Dynamical Systems. World Sci. Publishing, Singapore, pp. 287–290.
Ory, H., 1938. L'extraction des racines par la méthode heronienne. Mathesis 52, 229–246.
Páles, Z., 1988. Inequalities for sums of powers. J. Math. Anal. Appl. 131, 265–270.
Páles, Z., 1988a. Inequalities for differences of powers. J. Math. Anal. Appl. 131 (1), 271–281.
Pappus d'Alexandria, 1932. Collection Mathématique (trad. De Paul Ver Eecke). Librairie Blanchard, Paris.
Pasche, A., 1946. Application des moyens à l'extraction des racines carrés. Elem. Math. 1, 67–69.
Pasche, A., 1948. Queques applications des moyens. Elem. Math. 3, 59–61.
Peetre, J., 1989. Generalizing the arithmetic geometric means-a hopeless computer experiment. Internat. J. Math. Math. Sci. 12 (2), 235–245.
Peetre, J., 1991. Some observations on algorithms of the Gauss–Borchardt type. Proc. Edinburg Math. Soc. (2)34 (3), 415–431.
Phillips, G.M., 1981. Archimedes the numerical analyst. Amer. Math. Monthly 88 (3), 165–169.
Phillips, G.M., 1984. Archimedes and the complex plane. Amer. Math. Monthly 91 (2), 108–114.
Phillips, G.M., 2000. Two Millennia of Mathematics. From Archimedes to Gauss. CMS Books Math., vol. 6. Springer-Verlag, New York.
Pittenger, A.O., 1980. Inequalities between arithmetic and logarithmic means. Univ. Beograd. Publ. Elektro. Fak. 678–715, 15–18.
Pólya, G., Szegö, G., 1951. Isoperimetric Inequalities in Mathematical Physics. Ann. of Math. Stud., vol. 27. Princeton Univ. Press, Princeton, New York.
Raşa, I., Toader, G., 1990. On the rate of convergence of double sequences. Bulet. Şt. Inst. Politehn. Cluj-Napoca, Seria Matem. Apl. 33, 27–30.
Rassias, T.M., Kim, Y.-H., 2003. New characterizations of some mean values and functional equations. Mathematica 45(68) (2), 185–194.

Rassias, T.M., Kim, Y.-H., 2008. On certain mean value theorems. Math. Ineq. Appl. 11 (3), 431–441.

Reyssat, E., 1987. Approximation des moyennes aritmetico-géométriques. L'Enseign. Math. 33 (3–4), 175–181.

van de Riet, R.P., 1963. Some Remarks on the Arithmetical–Geometrical Mean. Tech. Note, TN35. Stichtung Math. Centrum, Amsterdam.

Rosenberg, L., 1966. The iteration of means. Math. Mag. 39, 58–62.

Salamin, E., 1976. Computation of π using arithmetic–geometric mean. Math. Comp. 30, 565–570.

Sándor, J., 1994. On Gauss' and Borchard's algorithms. Sem. Didactica Matem. (Cluj-Napoca) 10, 107–112 (in Romanian).

Sándor, J., 1995. On certain inequalities for means. J. Math. Anal. Appl. 189 (2), 602–606.

Sándor, J., 1996. On certain inequalities for means II. J. Math. Anal. Appl. 199, 629–635.

Sándor, J., 1997. On means generated by derivatives of functions. Int. J. Math. Educ. Sci. Technol. 28 (1), 146–148.

Sándor, J., 2001. On certain inequalities for means III. Arch. Math. 76, 34–40.

Sándor, J., Toader, G., 1990. On some exponential means. Babeş-Bolyai Univ. of Cluj-Napoca Preprint 7, 35–40.

Sándor, J., Toader, G., 1999. Some general means. Czechoslovak Math. J. 49, 53–62.

Sándor, J., Toader, G., 2002. On means generated by two positive functions. Octogon Math. Mag. 10 (1), 70–73.

Sándor, J., Toader, G., 2006. On some exponential means II. Internat. J. Math. Math. Sci.. 9 pp. http://downloads.hindawi.com/journals/ijmms/2006/051937.pdf.

Sándor, J., Toader, G., 2006a. A property of an exponential mean. Octogon Math. Mag. 14 (1), 253.

Sándor, J., Toader, G., Raşa, I., 1996. The construction of some means. General Math. 4, 63–71.

Schering, K., 1878. Zur Theorie der Borchardtsen arithmetisch-geometrisch Mittels aus vier Elementen. J. Reine Angew. Math. 85, 115–170.

Schlesinger, L., 1911. Uber Gauss' Jugendarbeiten zum arithmetisch-geometrischen Mittel. Deutsche Math. Ver. 20, 396–403.

Schoenberg, I.J., 1978. On the arithmetic–geometric mean. Delta 7, 49–65.

Schoenberg, I.J., 1982. Mathematical Time Exposures. The Math. Assoc. of America.

Schwab, J., 1813. Eléments de Géométrie. Première Partie, Nancy.

Seiffert, H.-J., 1993. Problem 887. Nieuw Arch. Wisk. 11, 176.

Seiffert, H.-J., 1995. Aufgabe 16. Wurzel 29, 87.

Seiffert, H.-J., 1995a. Ungleichungen für einen bestimmten Mittelwert. Nieuw Arch. Wisk. 13, 195–198.

Snell, W., 1621. Cyclometricus. Leyden.

Song, Y.-Q., Jiang, W.-D., Chu, Y.-M., Yan, D.-D., 2013. Optimal bounds for Toader mean in terms of arithmetic and contraharmonic means. J. Math. Inequal. 7 (4), 751–777.

Sternberg, W., 1919. Einige Sätze über Mittelwerte. Leipzig. Ge. Wiss. Ber. (Math.-Nat. Kl.) 71, 277–285.

Stieltjes, T.J., 1880. Notiz über einen elementaren Algoritmus. J. Reine Angew. Math. 89, 343–344.

Stirling, J., 1730. Methodus Differentialis. London.

Stolarsky, K.B., 1975. Generalizations of the logarithmic mean. Math. Mag. 48, 87–92.

Stolarsky, K.B., 1980. The power and generalized logarithmic means. Amer. Math. Monthly 87 (7), 545–548.

Stöhr, A., 1956. Neuer Beweis einer Formel über das reelle arithmetisch-geometrischen Mittel. Jber. Deutsch. Math.-Verein. 58, 73–79.

Sutô, O., 1914. Studies on some functional equations. I. Tôhoku Math. J. 6, 1–15.

Sutô, O., 1914a. Studies on some functional equations. II. Tôhoku Math. J. 6, 82–101.

Tietze, H., 1952–1953. Über eine Verallgemeinerung des Gausschen aritmetisch-geometrisch Mittel und die zugehörige Folge von Zahlen n-tupeln. Bayer. Akad. Wiss. Math.-Natur. Kl. S.-B., 191–195.

Toader, G., 1986. Complex generalized means. In: Primul Simpozion National "Realizări şi perspective în domeniul traductoarelor". II. Cluj-Napoca, pp. 46–49.

Toader, G., 1987. Generalized means and double sequences. Studia Univ. Babeş-Bolyai 32 (3), 77–80.

Toader, G., 1987a. Generalized double sequences. Anal. Numér. Théor. Approx. 16 (1), 81–85.

Toader, G., 1987b. On the rate of convergence of double sequences. Babeş-Bolyai Univ. of Cluj-Napoca Preprint 9, 123–128.

Toader, G., 1988. Mean value theorems and means. In: Proceedings of National Conf. Appl. Math. Mechan.. Cluj-Napoca, pp. 225–230.

Toader, G., 1988a. An exponential mean. Babeş-Bolyai Univ. of Cluj-Napoca Preprint 7, 51–54.

Toader, G., 1989. A generalization of geometric and harmonic means. Babeş-Bolyai Univ. of Cluj-Napoca Preprint 7, 21–28.

Toader, G., 1990. On bidimensional iteration algorithms. Babeş-Bolyai Univ. of Cluj-Napoca Preprint 6, 205–208.

Toader, G., 1990a. On the convergence of double sequences. Bulet. Stiint. Instit. Politeh. Cluj-Napoca, Matem. Apl. Mec. 33, 31–34.

Toader, G., 1991. Some remarks on means. Anal. Numér. Théor. Approx. 20, 97–109.

Toader, G., 1998. Some mean values related to the arithmetic–geometric mean. J. Math. Anal. Appl. 218, 358–368.

Toader, G., 1998a. The monotonicity of a family of means. Bull. Appl. & Comput. Math. 1465 (85A), 185–198.

Toader, G., 1998b. Some trigonometric means. Bull. Appl. & Comput. Math. 1562 (86A), 469–476.

Toader, G., 1998c. Some properties of means. Octogon Math. Mag. 6 (2), 14–19.

Toader, G., 1999. Seiffert type means. Nieuw Archief voor Wiskunde 17, 379–382.

Toader, G., 2000. Means obtained by composition. Automat. Comput. Appl. Math. 9 (1–2), 48–51.

Toader, G., 2001. On some generalized means of Kim. Acta Techn. Napocensis 44 (1), 107–112.

Toader, G., 2002. Integral generalized means. Math. Ineq. Appl. 5 (3), 511–516.

Toader, G., 2004. Complementariness with respect to Greek means. Automat. Comput. Appl. Math. 13 (1), 197–202.

Toader, G., 2005. Weighted Greek means. Octogon Math. Mag. 13 (2), 946–954.

Toader, G., 2007. Complementary means and double sequences. Annal. Acad. Paedag. Cracov., Studia Math. 6, 7–18. Folia 45. http://studmath.up.krakow.pl/index.php/studmath/article/view/42.

Toader, G., 2007a. Universal means. J. Math. Inequal. 1 (1), 57–64.

Toader, G., Costin, I., 2007. On invariance in the family of Stolarsky means. Automat. Comput. Appl. Math. 16 (2), 333–340.

Toader, G., Costin, I., Toader, S., 2012. Invariance in some families of means. In: Rassias, T.M., Brzdek, J. (Eds.), Functional Equations in Mathematical Analysis. In: Springer Optim. Appl., vol. 52, pp. 709–718.

Toader, G., Raşa, I., 1996. Some inequalities with integral means. General Math. 4, 73–82.

Toader, G., Rassias, T.M., 1999. New properties of some mean values. J. Math. Anal. Appl. 232, 376–383.

Toader, G., Sándor, J., 2006. Inequalities for general integral means. J. Ineq. Pure Appl. Math. 7 (1). Art. 13, 5 pp. https://www.emis.de/journals/JIPAM/images/141_05_JIPAM/141_05.pdf.

Toader, G., Toader, S., 2002. A partial operation with means. Automat. Comput. Appl. Math. 11 (2), 79–88.

Toader, G., Toader, S., 2002a. Integral generalized means of Gauss type. Automat. Comput. Appl. Math. 11 (1), 146–155.

Toader, G., Toader, S., 2003. A partial operation with power means. Automat. Comput. Appl. Math. 12 (1), 79–88.
Toader, G., Toader, S., 2005. Greek Means and the Arithmetic–Geometric Mean. RGMIA Monographs. Victoria University. 95 pp. http://rgmia.org/papers/monographs/Grec.pdf.
Toader, G., Toader, S., 2007. Means and generalized means. J. Ineq. Pure Appl. Math. 8 (2). Art. 45, 6 pp. https://www.emis.de/journals/JIPAM/images/152_07_JIPAM/152_07.pdf.
Toader, G., Toader, S., 2008. Some weighted geometric means. Automat. Comput. Appl. Math. 17 (3), 565–571.
Toader, G., Toader, S., 2010. Invariance of an extended logarithmic mean with respect to weighted Gini means. Autom. Comput. Appl. Math. 19 (2), 125–132.
Toader, S., 2002a. Derivatives of generalized means. Math. Ineq. Appl. 5 (3), 517–523.
Toader, S., Rassias, T.M., Toader, G., 2001. A Gauss type functional equation. Internat. J. Math. Math. Sci. 25 (9), 565–569.
Toader, S., Toader, G., 2002b. Greek means. Automat. Comput. Appl. Math. 11 (1), 159–165.
Toader, S., Toader, G., 2003a. Series expansion of Greek means. In: Proceedings of the International Symposium Specialization, Integration and Development, Section: Quantitative Economics. Babeş-Bolyai University Cluj-Napoca, Romania, pp. 441–448.
Toader, S., Toader, G., 2004. Complementary of a Greek mean with respect to another. Automat. Comput. Appl. Math. 13 (1), 203–208.
Toader, S., Toader, G., 2004a. Symmetries with Greek means. Creative Math. 13, 17–25.
Toader, S., Toader, G., 2004b. Generalized complementaries of Greek means. Pure Math. Appl. 15 (2–3), 335–342.
Toader, S., Toader, G., 2006. Complementary of a Greek mean with respect to Lehmer means. Automat. Comput. Appl. Math. 15 (1), 319–324.
Toader, S., Toader, G., 2006a. Operations with means. Pure Math. Appl. 17 (3–4), 511–518.
Toader, S., Toader, G., 2007a. Complementaries of Greek means with respect to Gini means. Int. J. Math. Appl. Stat. 11 (V07), 187–192.
Toader, S., Toader, G., 2007b. Complementaries with respect to Stolarsky means. Automat. Comput. Appl. Math. 16 (2), 341–348.
Todd, J., 1992. Solution of the Problem 91-17. SIAM Rev. 34 (4), 653–654.
Tonelli, L., 1910. Sull'iterazione. Giorn. Matem. Battaglini 48, 341–373.
Tricomi, F.G., 1965. Sull'algoritmo iterativo del Borchardt e su di una sua generalizzazione. Rend. Circ. Mat. Palermo (2) 14, 85–94.
Tricomi, F.G., 1965a. Un algoritmo iterativo per il calcolo della funzione seno. Rend. Accad. Naz. Lincei (8) 39, 146–150.
Tricomi, F.G., 1970. Sulle combinazioni lineari delle tre classiche medie. Atti Accad. Sci. Torino 104, 557–572.
Tricomi, F.G., 1975. Sugli algoritmi iterative nell'analisi numerica. Acad. Naz dei Lincei, 105–117. Anno CCLXXII.
Trif, T., 2006. Monotonicity, comparison and Minkowski's inequality for generalized Muirhead means in two variables. Mathematica 48(71) (1), 99–110.
Ume, J.S., Kim, Y.-H., 2000. Some mean values related to the quasi-arithmetic mean. J. Math. Anal. Appl. 252 (1), 167–176.
Uspensky, J.V., 1945. On the arithmetic–geometric mean of Gauss I, II, III. Math. Notae 5, 1–28, 57–88, 129–161.
Vamanamurthy, M.K., Vuorinen, M., 1994. Inequalities for means. J. Math. Anal. Appl. 183, 155–166.
Veinger, N.I., 1964. On convergence of Borchardt's algorithm. Zap. Gorn. Inst. Plehanov. 43, 26–32.
Vuorinen, M., 1998. Hypergeometric functions in geometric function theory. In: Srinivasa Rao, K., Jaganuthan, R., Van der Jeugy, G. (Eds.), Proceedings of Special Functions and Differential Equations. Allied Publishers.

Vythoulkas, D., 1949. Generalization of the Schwarz inequality. Bull. Soc. Math. Grèce 25, 118–127.

Wang, M.-K., Qiu, Y.-F., Chu, Y.-M., 2010. Sharp bounds for Seiffert means in terms of Lehmer means. J. Math. Inequal. 4 (4), 581–586.

Wimp, J., 1985. Multidimensional iteration algorithms. Rend. Sem. Mat. Univ. Politech. Torino, 319–334. Fasc. Sp.

Witkowski, A., 2013. On Seiffert-like means. https://arxiv.org/pdf/1309.1244v1.pdf.

Yang, Z.-H., 2012. Sharp bounds for the second Seiffert mean in terms of power means. http://arxiv.org/pdf/1206.5494v1.pdf.

Yang, Z.-H., 2012a. Sharp power means bounds for Neumann–Sándor mean. http://arxiv.org/pdf/1208.0895v1.pdf.

Yang, Z.-H., Song, Y.-Q., Chu, Y.-M., 2014. Sharp bounds for the arithmetic–geometric mean. J. Inequal. Appl. 2014, 192.

Zhuravskii, A.M., 1964. The aritmetico-geometric mean algorithm. Zapisk. Leningrad. Gorn. Inst. G.V. Plehanov 43 (3), 8–25.

List of Symbols

Subject Index

T

U

W

Author Index

T

U

V

W

Y

Z